Systems Engineering to Improve Traumatic Brain Injury Care in the Military Health System

WORKSHOP SUMMARY

David Butler, Jessica Buono, Frederick Erdtmann, and Proctor Reid, Editors

NATIONAL ACADEMY OF ENGINEERING *AND*
INSTITUTE OF MEDICINE
OF THE NATIONAL ACADEMIES

THE NATIONAL ACADEMIES PRESS
Washington, D.C.
www.nap.edu

THE NATIONAL ACADEMIES PRESS 500 Fifth Street, N.W. Washington, DC 20001

NOTICE: The project that is the subject of this report was approved by the Governing Board of the National Research Council, whose members are drawn from the councils of the National Academy of Science, the National Academy of Engineering, and the Institute of Medicine.

Support for this project was provided by the U.S. Army (Award No. W81XWH-07-P-0979). Any opinions, findings, or conclusions expressed in this publication are those of the workshop participants and do not necessarily reflect the view of the organization that provided support for the project.

International Standard Book Number-13: 978-0-309-12758-5
International Standard Book Number-10: 0-309-12758-0

Copies of this report are available from the National Academies Press, 500 Fifth Street, N.W., Lockbox 285, Washington, DC 20055; (888) 624-8373 or (202) 334-3313 (in the Washington metropolitan area); online at *http://www.nap.edu.*

For more information about the National Academy of Engineering, visit the NAE home page at **www.nae.edu**. For more information about the Institute of Medicine, visit the IOM home page at **www.iom.edu**.

Printed in the United States of America

THE NATIONAL ACADEMIES

Advisers to the Nation on Science, Engineering, and Medicine

The **National Academy of Sciences** is a private, nonprofit, self-perpetuating society of distinguished scholars engaged in scientific and engineering research, dedicated to the furtherance of science and technology and to their use for the general welfare. Upon the authority of the charter granted to it by the Congress in 1863, the Academy has a mandate that requires it to advise the federal government on scientific and technical matters. Dr. Ralph J. Cicerone is president of the National Academy of Sciences.

The **National Academy of Engineering** was established in 1964, under the charter of the National Academy of Sciences, as a parallel organization of outstanding engineers. It is autonomous in its administration and in the selection of its members, sharing with the National Academy of Sciences the responsibility for advising the federal government. The National Academy of Engineering also sponsors engineering programs aimed at meeting national needs, encourages education and research, and recognizes the superior achievements of engineers. Dr. Charles M. Vest is president of the National Academy of Engineering.

The **Institute of Medicine** was established in 1970 by the National Academy of Sciences to secure the services of eminent members of appropriate professions in the examination of policy matters pertaining to the health of the public. The Institute acts under the responsibility given to the National Academy of Sciences by its congressional charter to be an adviser to the federal government and, upon its own initiative, to identify issues of medical care, research, and education. Dr. Harvey V. Fineberg is president of the Institute of Medicine.

The **National Research Council** was organized by the National Academy of Sciences in 1916 to associate the broad community of science and technology with the Academy's purposes of furthering knowledge and advising the federal government. Functioning in accordance with general policies determined by the Academy, the Council has become the principal operating agency of both the National Academy of Sciences and the National Academy of Engineering in providing services to the government, the public, and the scientific and engineering communities. The Council is administered jointly by both Academies and the Institute of Medicine. Dr. Ralph J. Cicerone and Dr. Charles M. Vest are chair and vice chair, respectively, of the National Research Council.

www.national-academies.org

WORKSHOP STEERING COMMITTEE ON SYSTEMS ENGINEERING HEALTH CARE: TOOLS AND TECHNOLOGIES TO MAXIMIZE THE EFFECTIVENESS OF MEDICAL MISSION SUPPORT TO DOD

NORMAN R. AUGUSTINE (NAE) *co-chair*, Lockheed Martin Corporation (retired), Bethesda, Maryland
JEROME H. GROSSMAN (IOM) *co-chair*, Harvard University/ Kennedy School Health Care Delivery Project, Cambridge, Massachusetts (until April 2008)
DENIS CORTESE (IOM) *co-chair*, Mayo Clinic, Rochester, Minnesota (from April 2008)
SETH BONDER (NAE), The Bonder Group, Ann Arbor, Michigan
PATRICIA F. BRENNAN (IOM), College of Engineering, University of Wisconsin-Madison
THOMAS F. BUDINGER (NAE), University of California, Berkeley, and E.O. Lawrence Berkeley National Laboratory, Berkeley, California
BARRETT S. CALDWELL, Purdue University, West Lafayette, Indiana
MICHAEL P. DINNEEN, Military Health System, Washington, D.C.
PAUL M. HORN (NAE), New York University, New York City
MICHAEL S. JAFFEE, COL (s), Defense and Veterans Brain Injury Center, U.S. Air Force, Washington, D.C.
WILLIAM P. NASH, CAPT, U.S. Navy (ret), USMC/USN Liaison to the Defense Center of Excellence for Psychological Health and TBI, Burke, Virginia
ALEXANDER K. OMMAYA, Department of Veterans Affairs, Washington, D.C.
DAVID T. ORMAN, COL (ret), U.S. Army MEDCOM, Fort Sam Houston, Texas
RONALD POROPATICH, COL, Medical Corps, U.S. Army, Fort Detrick, Maryland
WILLIAM B. ROUSE (NAE), Georgia Institute of Technology, Atlanta, Georgia
NINA A. SAYER, Center for Chronic Disease Outcomes Research, Minneapolis, Minnesota

Project Staff

PROCTOR P. REID, director, Program Office, National Academy of Engineering

FREDERICK (RICK) ERDTMANN, director, Board on Military and Veterans Health, Institute of Medicine

DAVID BUTLER, senior program officer, Board on Military and Veterans Health, Institute of Medicine

JESSICA BUONO, research associate, Program Office, National Academy of Engineering

CAROL R. ARENBERG, managing editor, National Academy of Engineering

PENELOPE J. GIBBS, senior program associate, Program Office, National Academy of Engineering

PRISCILLA ARRIAGA, Anderson & Commonweal Intern, Program Office, National Academy of Engineering

This workshop summary is dedicated to the memory of **Jerome H. Grossman, M.D.**, a long-time member, friend, and leader in the National Academies and the primary motivator and intellectual compass for this workshop.

Preface

This workshop was the outcome of a sequence of events made possible by Dr. Jerry Grossman, who co-chaired a study in 2005 by the National Academy of Engineering (NAE) and Institute of Medicine (IOM) that culminated in the publication of *Building a Better Delivery System: A New Engineering/Health Care Partnership.* That report makes a strong case for taking advantage of the best of both disciplines—health care and operational systems engineering (a combination of science and mathematics to describe, analyze, plan, design, and integrate systems with complex interactions among people, processes, materials, equipment, and facilities)—to improve the efficiency and quality of health care delivery, as well as health care outcomes.

There is widespread agreement that the overall quality of health care delivered in the United States is not commensurate with the nation's high health care expenditures or its global leadership in advanced biomedical technologies, and reform of the nation's health care system is a high priority of government officials, caregivers, and patients. The premise of the NAE/IOM report is that there are lessons to be learned from the experiences of industries that have used operational systems engineering tools to make higher quality, less expensive products more efficiently. Dr. Grossman mounted a personal campaign to apply these ideas to move our health care system to a higher plane. Among those most interested in pursuing this approach are leaders in the U.S. Department of Defense (DOD) and Department of Veterans Affairs, who are committed to finding ways of improving the quality of care for military personnel, veterans, and their families.

Intrigued by the possibilities, DOD decided to sponsor a series of workshops to explore the potential of applying operational systems engineering principles and tools to military health care, beginning with the diagnosis and care of patients with traumatic brain injury (TBI), one of the most prevalent and challenging injuries suffered by warriors in Iraq and Afghanistan. TBI presents an extremely complex medical problem with a wide range of severity levels and presenting symptoms. TBI patients require coordinated, often prolonged care by people in many different specialties and organizations. In short, operational systems engineering tools have the potential to improve the care of these wounded warriors.

Workshops sponsored by the National Academies are intended to identify avenues for further exploration rather than to provide consensus findings or recommendations. The workshop summarized in this volume, "Harnessing Operational Systems Engineering to Improve Traumatic Brain Injury Care in the Military Health System," is a fitting memorial to Dr. Grossman who died suddenly during the planning stages of the workshop. We believe he would have celebrated this undertaking, not because it was held in honor of his memory, but because it demonstrates the potential for improvement in which he so passionately believed.

It is our hope that readers will be encouraged to explore the potential of applying systems engineering tools to improve health care delivery in their own areas of medicine.

Norman R. Augustine, *Co-chair*
Denis Cortese, *Co-chair*
Workshop Steering Committee on Systems Engineering Health Care: Tools and Technologies to Maximize the Effectiveness of Medical Mission Support to DOD

Acknowledgments

This report has been reviewed in draft form by individuals chosen for their diverse perspectives and technical expertise, in accordance with procedures approved by the National Research Council (NRC) Report Review Committee. The purpose of this independent review is to provide candid and critical comments that will assist the institution in making its published report as sound as possible and to ensure that the report meets institutional standards for objectivity, evidence, and responsiveness to the study charge. The review comments and draft manuscript remain confidential to protect the integrity of the deliberative process. We wish to thank the following individuals for their review of this report:

William B. Elliott, Quality and Operations Support, Kaiser Permanente
Elena Nightingale, Scholar-in-Residence, Institute of Medicine
Alexander K. Ommaya, Office of Research and Development, Department of Veterans Affairs
Ronald L. Rardin, Center for Innovation in Healthcare Logistics, Department of Industrial Engineering, University of Arkansas
Vinod K. Sahney, Blue Cross Blue Shield of Massachusetts
Judith L. Swain, Singapore Institute of Clinical Sciences, Agency for Science, Technology and Research, National University of Singapore, and University of California, San Diego

Although the reviewers listed above have provided many constructive comments and suggestions, they were not asked to endorse the final draft of the report before its release. The review of this report was overseen by Chris G. Whipple, ENVIRON. Appointed by the NRC, he was responsible for making certain that an independent examination of this report was carried out in accordance with institutional procedures and that all review comments were carefully considered. Responsibility for the final content of this report rests entirely with the authors and the institution.

Contents

Appendixes

Acronyms/Abbreviations

ACR	American College of Rheumatology
AFEB	Armed Forces Epidemiological Board
AFTH	Air Force Theater Hospital
AHLTA	Armed Forces Health Longitudinal Technology Application (formerly CHCS II), a DOD enterprise-wide electronic health record system A –*T* suffix specifies the in-theater component.
ANAM	Automated Neuropsychological Assessment Metric
BAS	Battalion Aid Station
BLI	blast lung injury
CA	cellular automata model
CASEVAC	casualty evacuation
CDC	Centers for Disease Control and Prevention
CDP	Center for Deployment Psychology
CDR	Clinical Data Repository
CONUS	Continental United States
CSTS	Center for the Study of Traumatic Stress
DARPA	Defense Advanced Research Projects Agency
DCOE	Defense Center of Excellence
DHCC	Deployment Health Clinical Center
DOD	U.S. Department of Defense
DVBIC	Defense and Veterans Brain Injury Center
DVHIP	Defense and Veterans Head Injury Program

EIA	enzyme immunoassay
EOD	Explosive Ordnance Disposal
EVAC	evacuation
FRSS	Forward Resuscitative Surgical System (USN/USMC)
FST	Forward Surgical Team (USA)
GAO	Government Accountability Office
GI	gastrointestinal
GPS	Global Positioning System
HIV	human immunodeficiency virus
HMMWV/ HUMVEE	High-Mobility Multipurpose Wheeled Vehicle
ICD	International Classification of Disease
ICU	intensive care unit
IED	improvised explosive device
IHI	Institute for Healthcare Improvement
IOM	Institute of Medicine
JTTS	Joint Theater Trauma System
MC4	Medical Communications for Combat Casualty Care
MCAS	Marine Corps Air Station
MDP	Markov decision process
MEDEVAC	medical evacuation
MHS	Military Health System
MIP	mixed-integer programming model
MRI	magnetic resonance imaging
MRP	material requirements planning
mTBI	mild traumatic brain injury
MTF	military treatment facility
NAE	National Academy of Engineering
NHLBI	National Heart Lung Blood Institute
NICoE	National Intrepid Center of Excellence for Psychological Health and Traumatic Brain Injury
OEF	Operation Enduring Freedom
OIF	Operation Iraqi Freedom
OSD/HA	Office of the Assistant Secretary of Defense for Health Affairs [also OSD(HA)]
OSE	operational systems engineering

OIPT	overarching integrated product team
PDHA	post-deployment health assessment
PDHRA	post-deployment health reassessment
PET	position emission tomography
POMDP	partially observable Markov decision process
PPOC	polytrauma points of contact
PRC	polytrauma rehabilitation center
PSCT	polytrauma support clinic team
PTSD	post-traumatic stress disorder
QALY	quality-adjusted life year
RMC	regional medical center
RPG	rocket-propelled grenade
RTD	return to duty
SDT	signal-detection theory
SPECT	single-photon emission-computed tomography
SSTP	Surgical Shock Trauma Platoon (USN/USMC)
TBI	traumatic brain injury
THA	total hip arthroplasty
TRAC^{2}ES	TRANSCOM Regulating and Command & Control Evacuation System
UHC	University Hospital Consortium
USAMRMC	U.S. Army Medical Research and Materiel Command
USTRANSCOM	U.S. Transportation Command (USAF)
VA	U.S. Department of Veterans Affairs
VHA	Veterans Health Administration
VHIS	Vietnam Head Injury Study
VISN 8	Veterans Integrated Services Network Region 8
VISTA	Veterans Health Information Systems and Technology Architecture
VSA	value-stream analysis
WB	western blot
WRAMC	Walter Reed Army Medical Center

Summary

The U.S. Department of Defense (DOD) vision for providing health care to the military and their families, spelled out in the *Roadmap for Medical Transformation*, is based on three "pillars"—high-quality health care delivery, in-house medical education, and research (MHS, 2006). DOD recognizes, however, that transformational advancements in the quality and productivity of the Military Health System (MHS) may require tools and techniques developed as part of operational systems engineering (OSE), a family of disciplines that includes industrial engineering, operations research, human factors engineering, and financial engineering/risk analysis, as well as from computer science and engineering and the social and behavioral sciences. All of these engineering and science disciplines are integrally involved in the design, analysis, and control of complex processes and systems.

Many companies in engineering-intensive manufacturing and service industries have also contributed to the development of OSE tools and methods as a result of their experiences with them in improving the performance of their own companies. The results of those applications of OSE have led to a comprehensive understanding of the function and dynamics of complex systems and insights into interactions between subsystems and processes in those industries. OSE tools can also support the rational, systematic management of the tensions and trade-offs between competing performance goals and priorities among stakeholders.

Building a Better Delivery System: A New Engineering/Health Care Partnership, a 2005 report by the National Academy of Engineering (NAE) and Institute of Medicine (IOM) of the National Academies, documented that the health care sector as a whole has been relatively slow to embrace OSE tools and techniques. These tools and techniques could help untangle the complexities and lead to a deeper understanding of the dynamics of health care systems and subsystems and could be designed to optimize system performance to meet specific quality goals (e.g., safety, patient-centeredness, timeliness) and, at the same time, improve prediction, measurement, and management to meet other performance goals (e.g., cost, access, productivity).

OSE, which combines science and mathematics to improve the operation of systems, has greatly benefited other enterprises by describing, analyzing, planning, designing, and integrating systems with complex interactions among people, processes, materials, equipment, and facilities using deterministic and probabilistic mathematics (called stochastic processes). The ultimate goal of OSE is to integrate all elements in the operations of a system to improve its efficiency and effectiveness.

The idea of applying the tools, techniques, and concepts of OSE to health care has been a topic of discussion for some years, and health care systems and care providers in some areas have adopted them successfully. In epidemiology, for example, OSE tools have been used to evaluate intervention strategies, disease-control programs, screening programs, and health-promotion and disease-prevention programs, and to predict the incidence, prevalence, and mortality rates of diseases. OSE concepts have also been used to design health care systems, estimate future resource needs, optimize the allocation of resources, and optimize capacity, plan facilities, and design emergency services in health care systems. In medical decision making, OSE techniques have been used to plan and implement appointment systems to reduce waiting time for both outpatients and inpatients, optimize staff levels and scheduling, conduct inventories, plan material requirements, optimize supply chains, forecast demand, plan auxiliary services, and evaluate medical technologies.

In 2007, the U.S. Army Medical Research and Materiel Command (USAMRMC) asked NAE and IOM to conduct a series of workshops to gather information and provide guidance to MHS on using OSE tools and technologies to improve the quality and productivity of health care delivery to the nation's armed forces and other eligible beneficiaries of TRICARE (military/civilian health care interfaces). As a first step,

BOX S-1
Department of Defense Definition of Traumatic Brain Injury

Traumatic brain injury (TBI) is a traumatically induced structural injury and/or physiological disruption of brain function as a result of an external force that is indicated by new onset or worsening of at least one of the following clinical signs, immediately following the event:

(1) Any period of loss, or a decreased level, of consciousness.
(2) Any loss of memory for events immediately before or after the injury.
(3) Any alteration in mental state at the time of the injury (confusion, disorientation, slowed thinking, etc.).
(4) Neurological deficits (weakness, loss of balance, change in vision, praxis, paresis/plegia, sensory loss, aphasia, etc.) that may or may not be transient.
(5) Intracranial lesion.

Source: DOD, 2007.

USAMRMC asked that a workshop be held to identify promising areas for near-, medium-, and long-term applications of OSE tools and information technologies for modeling, analyzing, designing, and improving the care and management of patients with traumatic brain injury (TBI) throughout the military health care continuum—from battlefield to field hospital to U.S.-based military health care facilities to TRICARE networks and U.S. Department of Veterans Affairs (VA) facilities (Box S-1). The expectation is that the application of OSE tools and techniques to TBI care will improve the delivery of care at both the tactical and strategic levels.

OSE tools could also be used to improve tracking and navigating of patients through critical transition points, such as from a health care organization in one military service to a facility in another, or to the VA or a civilian health care facility. Transition points, or handoffs, also occur when patients are moved from acute TBI care in a hospital to more chronic long-term management in a community clinic, or when there is a changeover in caregivers from one shift to another. Minimizing confusion, delays, and/or variations in the quality of care during hand-offs is a key area in which OSE techniques may be helpful.

During the planning phase of the workshop, a steering committee of experts in TBI, military and veterans health care delivery, and OSE, supported by NAE and IOM professional staff, compiled a list of issues raised by stakeholders in the MHS community related to the care of mild, moderate, and severe TBI cases.[1] From that list, they identified issues that could potentially benefit from OSE approaches and categorized them into five major challenge areas for TBI care:

(A) the development of new TBI knowledge
(B) the detection and screening of TBI conditions
(C) communication and coordination for TBI care
(D) measuring and forecasting the demand for TBI care
(E) the capacity, organization, and resource allocations of the TBI care system

The committee then condensed the major stakeholder issues in each of these categories into two or three issues for OSE analysis, that is, analytical challenges that could be met through the application of OSE approaches and that could lead to improvements in the delivery of TBI care.

This workshop summary synthesizes the results of a two-day NAE/IOM invitation-only workshop held in Washington, D.C., June 11–12, 2008. The introduction (Chapter 1) reviews the potential of OSE for improving the quality of health care and places the challenge of TBI care in the context of the broader issues of the quality of, and cost challenges to MHS health care. Chapters 2 through 5 provide individually authored summaries of the workshop presentations and discussions. These included background on the medical aspects of TBI and major clinical and logistical challenges in TBI care, as well as examples illustrating relevant applications of OSE tools and methods, and a case study of a shift by a major unit in an academic health system from expert-based medical practice to expert-managed system-supported practice.

Chapter 6 includes the charges to the five working groups that

[1]The level of severity is classically defined at the time of injury using measures of the state of consciousness (the Glasgow Coma Scale), the duration of the loss of consciousness, and post-traumatic amnesia.

were asked to address the five major challenge areas for TBI care and describes the results of their deliberations. It includes suggestions for 10 illustrative "analysis plans," that is, designs for potential OSE studies, analyses, and applications that could answer important questions raised by TBI stakeholders and help improve the performance of the TBI care delivery system. The suggestions for analysis plans were developed by the working groups to illustrate potential applications of OSE tools and methods to specific TBI care challenges. However, the working groups are not Academy-appointed committees, and the summaries reflect the views of the individuals who participated in each group discussion—and not necessarily the views of the institution or the workshop planning committee. The suggestions for OSE studies and projects should not be construed as consensus recommendations of the working groups, the workshop participants as a whole, or the National Academies.

SUMMARY OF SUGGESTIONS FOR ANALYSIS PLANS BY THE WORKING GROUPS

Figure S-1 captures the challenges raised during the TBI workshop from the perspective of OSE. As the figure shows, before OSE methods can be brought to bear, OSE practitioners must have a preliminary understanding of the relationships between blast and concussive injuries and TBIs. Working Group A was asked to address this issue, the development of TBI knowledge. To this end, the group developed an analysis plan for using diagnostic and screening tools to establish pre- and post-event baselines, as well as conducting basic research on blast and concussive effects.

The current MHS TBI care delivery system must be better specified and understood for OSE tools and methods to be effective. The complex military health care delivery system includes facilities, logistical support, and personnel in the MHS, VA, and civilian health care systems, as well as the families of soldiers suffering from TBI and the soldiers themselves. One of the basic challenges associated with the delivery of care is patient tracking and case management.

Working Group C participants suggested an approach to the development of an information system for tracking, monitoring, and cueing care delivery for all TBI patients. The approach focuses on the integration and augmentation of existing databases and a communications

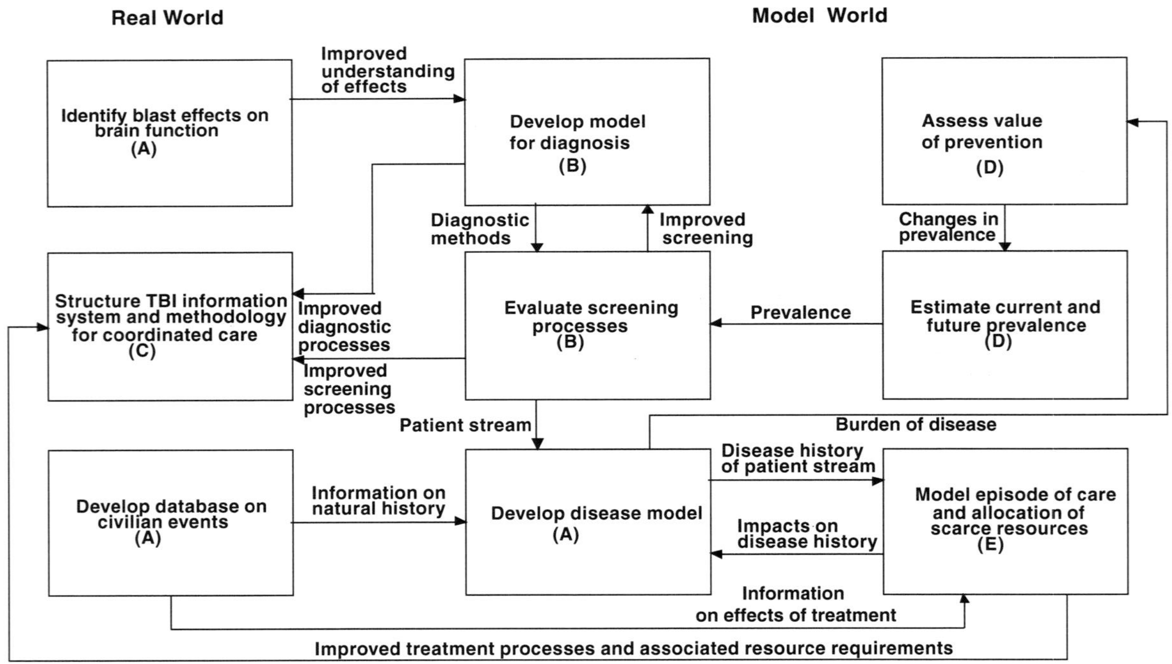

FIGURE S-1 Interrelationships among suggestions for analysis plans developed by participants in Working Groups A through E.

system that would provide access to, and the dissemination of, information to appropriate users. The architecture of the resulting system would be compatible with the military care delivery system.

During the workshop discussions, it was noted that data are also available from *outside* the military in the form of records of concussive and other closed-head injuries and TBI cases in the civilian world. These data could contribute to a more detailed understanding of the medical course of TBIs and the effectiveness of alternative treatments. Working Group A participants suggested an approach for integrating civilian and military data for use in the military environment.

The other blocks in Figure S-1 constitute typical, interrelated tasks that have been addressed by OSE techniques and methodologies in other contexts. The development of a quantitative disease model is essential to evaluating the long-term demand for care, strategies for treatment, and the design (and costs) of a high-quality, efficient care delivery system. Working Group A participants developed an approach that would model the course of TBI cases using a finite state-space stochastic process in which patients transition from one state to another as a function of additional (non-TBI) trauma, treatment, the long-term impacts of trauma, and co-morbidities that impact the state definition and occupancy times in any given state. The group also developed a plan for using a survey methodology integrated with data mining to define states, estimate transition probabilities, and determine the distributions of occupancy times.

Parameters in the quantitative disease model include diagnosis and screening characteristics, state definitions, and state transitions. Working Group B participants outlined the development of a series of models to quantitatively describe and evaluate current practices and to then optimize the screening process based on an analysis of data obtained from data mining and a survey. The group noted that Markov decision theory, Bayesian networks, influence diagrams, and simulations would be viable approaches to evaluating and designing TBI diagnostic and screening processes.

For MHS to assess its capacity to treat TBI, it must be able to understand and predict the "demand" on the system. Working Group D participants observed that estimating demand is extremely difficult, complicated by the many instances of TBI, especially mild TBI (mTBI), that are not reported promptly in theater or are unrecognized in post-deployment interviews. Group D participants observed that historical data could be analyzed as a basis for making statistical estimates of the number of mTBI cases in the current population of military personnel who have served or are serving in Iraq and Afghanistan. The group then outlined a methodology for forecasting future mTBI cases, taking into account conditions in the military theater, threats, and role-based exposures. The group also addressed the challenge of assessing the value of TBI prevention efforts, enlisting methods of estimating which specific investments in prevention strategies might yield reductions in TBI incidence, as well as cost savings to the health care system and reduction in the burden on soldiers and their families.

Given a quantitative disease model and an understanding of the effectiveness, availability, and costs of various treatments, it is possible to design and evaluate care delivery from the perspective of individual patients, a patient population, and the entire health care enterprise. Working Group E participants designed an approach to the development of an enterprise-level health care delivery model that would address, quantitatively, a broad spectrum of TBI treatment capacity, organizational, and resource allocation issues that would support decisions on policy and the design of the health care enterprise.

The suggestions for analysis plans developed by the five working groups indicate how OSE methods and tools could contribute to meeting the challenges of delivering effective, efficient, high-quality TBI care and management. Particular quantitative methods and models could provide insights into the design of diagnostic and screening processes, the delivery of care, the sizing of facilities, and the design of the overall health care delivery complex to meet current and future demands. As shown in Figure S-1, the challenges addressed by the OSE analysis plans are interrelated with the outputs of the plans. Each plan could potentially provide inputs to the development of one or more of the other plans. Collectively, the plans address many of the issues initially raised by TBI stakeholders.

The suggestions and assumptions of the working groups revealed several common interdependencies that are important to the development and application of OSE methods and tools to TBI care. These dependencies cut across the five focus areas:

- an assumption that sufficient, reliable data are available for the development of initial versions of all of the approaches outlined by the working groups and that additional reliable data could be

generated to improve and refine these approaches and provide more comprehensive, precise support to meet MHS needs in providing TBI care
- the value of standardized, detailed coding to improve the accuracy of records of TBI symptoms, injuries, and, possibly, treatments
- the importance of data sharing and interoperability of databases relevant to TBI diagnosis, treatment, measurement, and prediction
- the need for extensive, detailed maps of care paths and processes and associated information, patient, provider, and material flows for TBI care

The ideas and concepts introduced during the workshop may be helpful to DOD leaders working to refine and improve DOD's system of health care delivery, both at the individual patient-provider level and at the enterprise level. Applications of OSE concepts, tools, and methods may potentially contribute to improvements in care, not only for TBI patients but for all patients receiving MHS health care.

REFERENCES

DOD (U.S. Department of Defense). 2007. Traumatic Brain Injury: Definition and Reporting. Memorandum. HA Policy 07-030. Dated October 1, 2007. Available online at *http://mhs.osd.mil/Content/docs/pdfs/policies/2007/07-030.pdf* (accessed August 13, 2008).

MHS (Military Health System). 2006. QDR Medical Roadmap Implementation. Military Health System Office of Transformation PPT Presentation, July 13, 2006. Available online at *http://www.tricare.osd.mil/ocfo/_docs/071306_qdr_trans_infra.ppt#0* (accessed September 11, 2007).

NAE/IOM (National Academy of Engineering/Institute of Medicine). 2005. Building a Better Delivery System: A New Engineering/Health Care Partnership, edited by P.P. Reid, W.D. Compton, J.H. Grossman, and G. Fanjiang. Washington, D.C.: The National Academies Press.

1

Introduction

Engineers have long been closely involved in the development of medical technologies (devices, equipment, pharmaceuticals) and in supporting medical research (instrumentation, computational tools, etc.) (IOM, 1995; NAE, 2003). Revolutionary advances in bioengineering and genomics and the promise of quantum advances in diagnostic tools and therapies testify to the vitality of the partnership.

Yet engineers and engineering techniques have remained on the periphery of efforts to understand, assess, and manage or redress the challenges involved in health care delivery. Even information and communications technologies, which have been widely applied to the administrative and financial aspects of the health care industry, have had relatively little impact on the core business of health care—clinical operations. Moreover, the principles, tools, and research of an entire "family" of engineering disciplines associated with the analysis, design, and control of complex systems (operational systems engineering [OSE], which includes aspects of industrial engineering, operations research, human factors engineering, and financial engineering/risk analysis) have largely been absent from the clinical operations of health care delivery.

OSE combines science and mathematics to improve the operations of systems and enterprises that provide goods and services by describing, analyzing, planning, designing, and integrating systems with complex interactions among people, processes, materials, equipment, and facilities. Operational systems engineers use deterministic and probabilistic mathematics (called stochastic processes) to describe how a system operates, change the design of a system based on those descriptions, and integrate all elements of operations, including people, processes, materials, and equipment, to improve the efficiency and effectiveness of the system at all levels.

The multiple crises facing our health care system related to quality, cost, and access clearly reflect the complexity of the current system and the urgent need for information and OSE tools that can help unravel some of those complexities and ultimately improve the quality of care. The health care system must begin to analyze and resolve some of the difficult tensions and trade-offs among the six areas of urgent need identified by the Institute of Medicine (IOM, 2001)—safety, effectiveness, efficiency, timeliness, patient-centeredness, and equity. In addition, the analysis must take into account the competing objectives, priorities, and quality perspectives of patients, physicians, nurses, administrators, insurers, regulators, and other "stakeholders."

These kinds of complex trade-offs are not unique to health care. Product manufacturers (e.g., automakers) also make trade-offs between product features, for example, that might reduce maintenance costs but increase manufacturing costs and, in turn, increase the retail cost of the product. Many engineering-intensive manufacturing and service industries have worked with operation systems engineers to analyze trade-offs and otherwise improve the efficiency of their operations. Over a period of decades, these experiences have demonstrated the value of OSE tools and methods in deepening the understanding of the function and dynamics of complex systems, providing insights into interactions between subsystems and processes, and, ultimately, enabling more effective management and more efficient performance.

In the 2005 National Academy of Engineering (NAE)/IOM report, *Building a Better Delivery System: A New Engineering/Health Care Partnership*, these same engineering tools were shown to have the potential of doing the same for the delivery and management of health care. However, although OSE seems a natural partner for addressing some of the challenges facing the health care system, practitioners of the two disciplines are still largely ignorant of each other's methods, metrics, values, and mindsets. Most clinicians and health care administrators have had little exposure to the problem-solving methodologies and vocabulary of engineers, and few engineers are knowledgeable about the complex sociotechnical fabric of health care processes and systems. Thus neither has been able to communicate with the other in terms that would lead to fruitful collaboration to address the growing crises of health care delivery (NAE/IOM, 2005).

HEALTH CARE QUALITY AND COST CHALLENGES FACING THE MILITARY HEALTH SYSTEM

The U.S. Military Health System (MHS) is a $40 billion per year enterprise that provides operational medicine, training, research, and Force Health Protection support across the full range of military operations. MHS delivers health care services to 9.2 million eligible beneficiaries through TRICARE direct and managed care programs and directly controls an extensive integrated health care delivery system encompassing salaried health care personnel, facilities, infrastructure, research, education and training capabilities, and other assets.

Roughly 70 percent of all care received by TRICARE beneficiaries is purchased by participating private-sector health care providers. Collectively, the TRICARE benefit includes a national network of more than 220,000 physicians, all U.S. hospitals, and approximately 55,000 retail pharmacies (TRICARE, 2006). It is estimated that about a third of TRICARE beneficiaries will be served by its care programs for 50 years or more.

Like the rest of the nation's health care delivery enterprise in which it is embedded, MHS faces a number of pressing challenges related to the quality and cost of health care. Although MHS is committed to making evidence-based medicine the standard throughout the TRICARE system, the diffusion and application of best-practice treatments for illnesses and recommended processes for care pose significant challenges to the system.

Like health care costs in the overall U.S. economy, the costs of defense health programs are increasing much faster than inflation and the U.S. Department of Defense (DOD) budget overall. TRICARE costs more than doubled, from $19 billion to $38 billion, in the five-year period from FY2001 to FY2006, and they are projected to reach $64 billion by FY2015 (TRICARE, 2006). The sheer scope, diversity, and highly distributed nature of the MHS integrated care delivery system, which extends from battlefields and field hospitals overseas to U.S.-based military treatment facilities to a large, diverse assemblage of partnering (contracting) U.S.-based private-sector health care providers, make the tasks of improving quality and efficiency of care delivery that much more difficult. Additional burdens have been placed on the system by the needs of members of the armed forces and National Guard/Army Reserve who have sustained injuries in Iraq and Afghanistan.

THE OPERATIONAL SYSTEMS ENGINEERING IMPERATIVE FOR THE MILITARY HEALTH SYSTEM

In many respects, MHS is well positioned to meet the challenges described above. An extensive "in-house" integrated delivery system, in which physicians and other health professionals are salaried employees organized into multispecialty group practices governed by strong managerial hierarchies, has demonstrated a capacity for introducing and carrying through system-wide changes in the organization and the delivery of care.

Over the past decade, DOD has invested heavily in health information infrastructure to support its operations. The backbone of this infrastructure is the Armed Forces Health Longitudinal Technology Application (AHLTA), an extensive electronic health record that combines clinical and other health-related information from patient encounters worldwide. AHLTA currently supports the delivery of medical care to more than 7 million of its 9.2 million beneficiaries. AHLTA captures more than 55,000 patient encounters every workday and contains data on more than 9.5 million outpatient encounters.

DOD is working closely with the Department of Veterans Affairs (VA) to bring about full interoperability of AHLTA with the Veterans Health Information Systems and Technology Architecture (VISTA), the VA's highly acclaimed electronic health record. Integration of the two systems will provide a seamless exchange of medical information while continuing to serve the distinct core missions of each (Versel, 2006). Although not yet fully realized, AHLTA and its underlying information systems have the potential for supporting both evidence-based medicine and evidence-based management throughout MHS—a transition that will be critical to improving the quality and reducing the costs of military health care.

MHS has also been more aggressive than most other health care providers in adopting proven business planning and quality improvement methods from the private sector (e.g., Six-Sigma method, Toyota Production System, Continuous Process Improvement concepts and tools) to improve its performance (DOD, 2006). In the *Quadrennial Defense Review: Roadmap for Medical Transformation*, MHS laid out an expansive agenda for transforming its health professional workforce, infrastructure, clinical and business operations, financing, and the participation of TRICARE beneficiaries in their own health care (MHS, 2006).

MHS also has an extensive in-house education, training, research, and testing infrastructure, including its own medical school (Uniformed Services University of the Health Sciences), and numerous research centers (e.g., National Capital Area Simulation Center, Center for Education and Research in Patient Safety, and others) that have been enlisted fully in the Medical Transformation agenda.

Even though MHS has laid a foundation for addressing the massive health care quality and cost challenges ahead, there has been a growing awareness in DOD that building on this foundation and bringing about a sustainable transformation in the quality and productivity of its operations will require that MHS draw on a wide array of tools, techniques, technology, and knowledge developed in the disciplines of systems engineering, industrial engineering, operations research, human factors, computer science and engineering, and the social and behavioral sciences for the design, analysis, and control of complex processes and systems. These tools and techniques have also been improved through experience in their applications to many engineering-intensive manufacturing and service industries.

These same engineering tools and technologies could help MHS optimize system performance with respect to specific goals, such as safety, patient-centeredness, and timeliness, and improve its anticipation, measurement, and management of the effects of these interventions on other performance goals, such as cost, access, and productivity. Nevertheless, the health care sector as a whole, including MHS, has been relatively slow to embrace OSE tools and techniques (NAE/IOM, 2005).

There have been some interactions and successes, however. Small numbers of clinicians, care teams, and administrators, most of them based in large, integrated, salaried, multispecialty group practices, have adapted some OSE tools, such as statistical process control, queuing theory, and human factors engineering, to health care on a tactical level to improve the performance of discrete care processes, units, and departments. Unfortunately, to date little strategic use has been made of more information- and information technology (IT)-intensive OSE tools and techniques for analyzing and optimizing performance at higher levels of the health care system, such as for individual health care organizations, regional care systems, and the public health system (NAE/IOM, 2005).

Moreover, information about the successes, failures, and lessons learned by these early adopters of systems tools and techniques has not been systematically gathered or widely shared. In addition, although economic, organizational, managerial, educational, policy-related, and other barriers to the use of systems tools and complementary knowledge have been characterized and some strategies for overcoming them have been developed, this information has not been widely disseminated.

GOALS OF THE WORKSHOP

In 2007, the U.S. Army Medical Research and Materiel Command (USAMRMC) asked NAE and IOM to conduct a series of workshops to gather information and provide guidance to MHS on using systems tools and technologies to improve the quality and productivity of health care delivery to the nation's armed forces and other eligible beneficiaries of TRICARE. As the initial activity, USAMRMC asked the National Academies to undertake a workshop using a case-study approach to identify promising areas for the near-, medium-, and long-term application of OSE tools and IT to the modeling, analysis, design, and improvement of the care of patients with traumatic brain injury (TBI) across the military health care continuum—from battlefield to field hospital and U.S.-based military health care facilities to TRICARE networks (defense/civilian health care interfaces) and VA health care facilities. The ultimate objective is to improve the delivery of TBI care at both the tactical and strategic levels.

During the planning phase of the workshop, a steering committee of experts in TBI, military and veterans health care delivery, and OSE (see Appendix A), supported by NAE and IOM professional staff, compiled a list of issues raised by the MHS community related to the care of mild, moderate, and severe TBI cases (see Appendix B). This extended list of stakeholder issues was drawn from recent studies and reports on MHS care of TBI cases[1] and a preliminary meeting of steering committee

---

[1]See references to Chapters 2 and 3.

members and NAE and IOM staff with representatives of MHS on December 20, 2007 (see Appendix C). At the workshop planning meeting in February 2008, the committee identified a subset of issues from this list that could potentially benefit from OSE (see Appendix D). Chapter 6 details how these issues were translated into focus areas and discussion topics for the workshop (see Appendix E).

THE CHALLENGE OF TRAUMATIC BRAIN INJURY CARE

TBI—often called the signature injury of the Iraq and Afghanistan conflicts—is caused by a blow, jolt, or penetrating wound to the head that disrupts the normal functioning of the brain. In fact, TBIs account for nearly one-third of the injuries incurred by soldiers evacuated to Walter Reed Army Medical Center from the Operation Iraqi Freedom/Operation Enduring Freedom (OIF/OEF) theater. Between January 2003 and March 2008, more than 6,600 patients with TBIs were seen at MHS, VA, and civilian hospitals responsible for providing care under the aegis of the Defense and Veterans Brain Injury Center (DVBIC, 2008).

TBI is not only a wartime injury, however. The Centers for Disease Control and Prevention estimate that at least 1.4 million people in the United States incur TBIs each year, primarily as a result of falls, motor vehicle accidents, assaults, or being struck by or against an object[2] (CDC, 2006). Of these, about 235,000 are hospitalized, and about 50,000 die. However, even these numbers, troubling as they are, probably underestimate the problem.

Mild TBI (mTBI), which accounts for the vast majority of cases, is poorly documented and sometimes goes unrecognized in the presence of other injuries or does not become apparent for a period of time after the incident. Nonetheless, the neurologic and cognitive impacts of mTBI—which can include impaired memory and attention/concentration, headaches, slowed thinking, irritability, depression, and sleep disturbances (DVBIC, 2008)—can profoundly affect the lives of the injured and their families.

According to a 2008 report by the Government Accountability Office, caregivers in the MHS[3] treating patients with TBI, especially

---

[2]TBIs resulting from sports injuries often fall into this category.

[3]The MHS is a DOD enterprise comprising "the Office of the Assistant Secretary of

mTBI, face substantial challenges: (1) no objective diagnostic tests for TBI are available; (2) the symptoms of mTBI overlap with those of other disorders and with health complaints common in the general population; and (3) OIF/OEF veterans with mTBI may not realize that their health problems are associated with TBI and thus may not seek care (GAO, 2008). The report also suggests that military personnel may be reluctant to acknowledge the condition, because it might be perceived (wrongly) as a mental illness or might affect their military careers.

In a 2006 memo to the Assistant Secretary of Defense for Health Affairs, the Armed Forces Epidemiological Board identified several gaps in knowledge related to TBI and recommended measures to address those gaps (AFEB, 2006). Among the latter were getting a better understanding of blast-associated TBI; developing a "DoD-wide consensus of care," including standardized methods of battlefield assessment of TBI and follow-up clinical evaluations; developing requirements for more efficient and effective documentation of injuries and their disposition; and educating service members, their families, and "any individual in a position to encounter and care for soldiers at risk for a TBI during or after military service." As the number of OIF/OEF veterans with brain injuries increases and as screening and evaluation programs expand, the demands on the TBI care delivery system will almost certainly increase.

ORGANIZATION OF THE WORKSHOP SUMMARY

The remainder of this workshop summary provides a discussion of the problems and challenges of TBI for MHS, the promise of OSE analysis as a tool for improving the understanding of TBI and the delivery of care, and the results of the NAE/IOM workshop. Chapters 2 through 5 summarize the presentations and discussion during the opening plenary session, which provided background on the medical aspects of TBI and major clinical and logistical challenges to TBI care and included illustrative examples of applications of OSE tools and

---

Defense for Health Affairs; the medical departments of the Army, Navy, Marine Corps, Air Force, Coast Guard, and Joint Chiefs of Staff; the Combatant Command surgeons; and TRICARE providers (including private sector health care providers, hospitals, and pharmacies)" (MHS, 2008). TRICARE defines itself as "the Department of Defense's worldwide health care program for uniformed service members and their families" (TRICARE, 2008).

techniques relevant to TBI care management, including a case study of a unit in a major academic health system that shifted from expert-based medical practice to expert-managed system-supported practice. Chapter 6 presents the charges to the working groups and the results of their deliberations, specifically 10 suggestions for illustrative "analysis plans" (i.e., designs for potential OSE studies/analyses/applications that could answer important questions raised by TBI stakeholders and ultimately improve the performance of the TBI care delivery system).

REFERENCES

AFEB (Armed Forces Epidemiological Board). 2006. Traumatic Brain Injury in Military Service Members, 2006-02. Memorandum for the Honorable William Winkenwerder, Jr., M.D., Assistant Secretary of Defense for Health Affairs. Dated August 11, 2006. Available online at *http://www.ha.osd.mil/afeb/2006/2006-02.pdf* (accessed August 20, 2008).

CDC (Centers for Disease Control and Prevention). 2006. Traumatic Brain Injury in the United States: Emergency Department Visits, Hospitalizations, and Deaths. Prepared by Division of Injury Response, National Center for Injury Prevention and Control, Centers for Disease Control and Prevention, U.S. Department of Health and Human Services. January 2006. Available online at *http://www.cdc.gov/ncipc/pub-res/TBI_in_US_04/TBI%20in%20the%20US_Jan_2006.pdf* (accessed August 18, 2008).

DVBIC (Defense and Veterans Brain Injury Center). 2008. Blast Injury FAQs. Available online at *http://www.dvbic.org/cms.php?p=Blast_injury* (accessed August 19, 2008).

DOD (U.S. Department of Defense). 2006. Continuous Process Improvement Transformation Guidebook, DTD May 06. Washington, D.C.: Department of Defense. Available online at *https://acc.dau.mil/CommunityBrowser.aspx?id=32364* (accessed September 10, 2008).

GAO (Government Accountability Office). 2008. VA Health Care: Mild Traumatic Brain Injury Screening and Evaluation Implemented for OEF/OIF Veterans, but Challenges Remain. GAO-08-276. Available online at *http://www.gao.gov/new.items/d08276.pdf* (accessed 19 August 2008).

IOM (Institute of Medicine). 1995. Sources of Medical Technology: Universities and Industry, edited by N. Rosenberg, A.C. Gelijns, and H. Dawkins. Washington, D.C.: National Academy Press.

IOM. 2001. Crossing the Quality Chasm: A New Health System for the 21st Century. Washington, D.C.: National Academy Press.

MHS (Military Health System). 2006. Quadrennial Defense Review: Roadmap for Medical Transformation. Military Health System Office of Transformation PowerPoint Presentation, July 13, 2006. Available online at *http://www.tricare.osd.mil/ocfo/_docs/071306_qdr_trans_infra.ppt#0* (accessed September 10, 2008).

MHS. 2008. What is the MHS? Available online at *http://www.health.mil/aboutMHS.aspx* (accessed August 20, 2008).

NAE (National Academy of Engineering). 2003. The Impact of Academic Research on Industrial Performance. Washington, D.C.: The National Academies Press.

NAE/IOM (Institute of Medicine). 2005. Building a Better Delivery System: A New Engineering/Health Care Partnership, edited by P.P. Reid, W.D. Compton, J.H. Grossman, and G. Fanjiang. Washington, D.C.: The National Academies Press.

TRICARE. 2006. Sustaining Your Military Health Care Home Page. Available online at *http://www.tricare.osd.mil/STB/index.cfm* (accessed February 13, 2006).

TRICARE. 2008. What is TRICARE? Available online at *http://www.tricare.mil/mybenefit/home/overview/WhatIsTRICARE?* (accessed August 20, 2008).

Versel, N. 2006. EHR offensive. For the Record 18(4): 26. Available online at *http://www.fortherecordmag.com/archives/ftr_02202006p26.shtml.*

2

Medical Aspects of Traumatic Brain Injury[1]

Robert Labutta

Between January 2003 and April 2006, 28 percent of all combat casualties from Operation Iraqi Freedom/Operation Enduring Freedom requiring treatment at Walter Reed Army Medical Center had a traumatic brain injury (TBI). In a series of TBI cases treated at Walter Reed, 68 percent were caused by blast events (Warden, 2006). Thus understanding TBI, especially some aspects of TBI, such as how blasts cause such injuries, is crucial to military medicine. The topic has been addressed at length in a number of publications (Hoge et al., 2008; Okie, 2005, 2006; RAND, 2008; Singer, 2008), and the summary that follows is intended to provide a general understanding of TBI to set the stage for the subsequent discussion of how OSE technology and techniques might improve the delivery of TBI care.

DEFINITION AND CATEGORIZATION OF TRAUMATIC BRAIN INJURY

The evolving definition of TBI, as currently used by the military, is shown in Box 2-1. Note that TBI involves both an injurious incident

[1]This chapter is based on the author's presentation and responses to questions raised during the plenary session of the NAE-IOM workshop on Harnessing Systems Engineering to Improve Traumatic Brain Injury Care in the Military Health System on June 11, 2008. The author would like to thank IOM staff member Rick Erdtmann for his assistance in preparing this material for publication.

BOX 2-1
Department of Defense Definition of Traumatic Brain Injury

Traumatic brain injury (TBI) is a traumatically induced structural injury and/or physiological disruption of brain function as a result of an external force that is indicated by new onset or worsening of at least one of the following clinical signs, immediately following the event:

(1) Any period of loss, or a decreased level, of consciousness.
(2) Any loss of memory for events immediately before or after the injury.
(3) Any alteration in mental state at the time of the injury (confusion, disorientation, slowed thinking, etc.).
(4) Neurological deficits (weakness, loss of balance, change in vision, praxis, paresis/plegia, sensory loss, aphasia, etc.) that may or may not be transient.
(5) Intracranial lesion.

Source: DOD, 2007.

and a change in mental status. The lower limit for closed TBI does not require a loss of consciousness.

TBI can be categorized in several ways, such as by the mechanism of the injury: (1) closed-head injuries caused by blunt-force trauma, such as a blow to the head from a fist, a fall, or a car crash; and (2) open-head injuries caused by a penetrating object, such as a bullet, shrapnel, or debris. Blast injuries may involve both categories. TBI can also be categorized by the area(s) of the brain affected, the extent of the effects (diffuse or localized brain involvement), and the duration of effects immediately after the injury.

The most common way to categorize non-penetrating TBI is by the severity of the injury—mild, moderate, or severe. The level of severity is classically defined at the time of injury using measures of the state of consciousness (the Glasgow Coma Scale), the duration of the loss of consciousness, and post-traumatic amnesia. The criteria that define these levels are listed in Table 2-1.

TABLE 2-1 Severity of TBI Based on Clinical Signs

Severity	Glasgow Coma Scale	Loss of Consciousness	Post-Traumatic Amnesia
Mild	13–15	< 20 min–1 hr*	< 24 hr
Moderate	9–12	1–24 hrs	> 24 hrs < 7 days
Severe	3–8	> 24 hrs	> 7 days

*This is the range for the upper limit; the lower limit is defined as any alteration in mental status. Source: Adapted from Helmick et al., 2007.

In some clinical situations, however, the initial categorization of a TBI does not necessarily correspond to the eventual extent of neurological damage or the eventual outcome. For example, the initial injury may not be categorized as severe, but if there is subsequent significant brain swelling or bleeding in the head (intracerebrally, under the dura mater, or between the dura mater and the skull), the pressure created may cause the patient to develop clinically severe dysfunction and marked disability.

TBIs are quite common and should not be construed as injuries that occur principally or only during combat. Many TBIs occur in civilian populations and in military populations during peacetime. The CDC estimates that there are 50,000 deaths, 235,000 hospitalizations, and more than 1 million emergency room visits each year due to TBIs (CDC, 2006). Although there are a number of mild, moderate, severe, and penetrating combat-related TBIs, many due to blast injuries, the number of non-combat injuries, such as injuries from motor vehicle crashes, assaults, and falls, remains a significant health care issue for the military. The number of mild TBIs (mTBIs) in military and civilian populations must be estimated because many people with mTBI in both populations do not receive medical evaluation. Langlois and colleagues (2006), for example, calculate that 1.6 to 3.8 million sports-related concussions that are not included in the CDC numbers above occur in the United States annually. Concussion and mTBI are synonyms and can be used interchangeably.

TRAUMATIC BRAIN INJURY IN THE MILITARY ENVIRONMENT

The recent increase in combat-related TBIs is largely because of a striking number of mTBI cases. Figure 2-1 shows the number of military

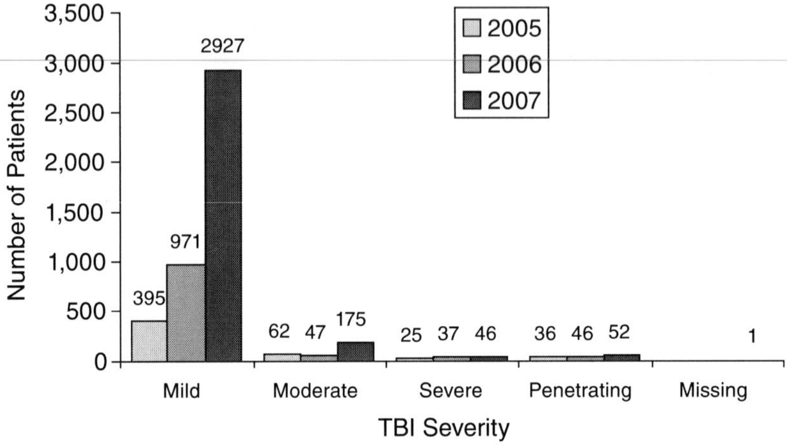

FIGURE 2-1 Severity of TBI cases treated at DVBIC Sites, 2005–2007. Source: Labutta, 2008.

personnel with TBI war wounds (by severity level) who were evacuated out of the combat theater to one of the DVBIC network sites between 2005 and 2007. In other words, moderate, severe, and penetrating TBI are easily recognized, recorded, and accounted for. In addition, the effects on the individual are usually obvious and generally result in long-term disability. The number of these injuries is directly related to the intensity of conflict and type of weaponry. However, for mTBI, also known as concussion, the number recorded is primarily a function of patient reporting and provider identification, since the effects of mTBI are less evident by comparison, are most often transient, and may be overshadowed by other diagnoses and mission requirements. Consequently, the disproportionate part of the increase in mTBI during 2007 may the result of improved processes and an increase in the locations where cases are ascertained and reported.

Among deployed U.S. military personnel, TBIs—especially moderate to severe cases—are often associated with other injuries to the face, neck, spine, and extremities, particularly following a blast. Improved body armor and advanced lifesaving measures—including new equipment and drugs, better treatment protocols, and faster transport to higher level medical assistance—have resulted in more troops surviving serious wounds. Awareness of TBI among health professionals has also increased, which has resulted in more intensive screening and detection and fewer misdiagnoses.

DIAGNOSIS AND TREATMENT

In patients with multiple injuries, TBI may not be clinically apparent unless there is a penetrating wound to the head. Even if TBI is suspected, it may not be the most urgent clinical need (e.g., a patient with mTBI may also have massive hemorrhaging due to an extremity injury). In these situations, the urgency and the priority of care may be directed toward other wounds and their overall effects. In general, for moderate, severe, and penetrating TBI, clinical care is directed toward the prevention of brain swelling, bleeding in the head, and poor cerebral blood flow. These are the secondary effects of TBI that can lead to further neuronal damage. As such, hemorrhage, hypotension, hypoxia, electrolyte imbalance, fever, seizures, and infection must receive initial attention and control.

The early identification and management of subdural hematoma is critical. The mortality rate associated with a four-hour delay following injury is 85 percent, compared with 30 percent if there is surgical control within four hours of the injury (Seelig et al., 1981). Unchecked brain swelling following an injury is extremely serious. In many cases, a large portion of the skull may be removed to relieve intracranial pressure and allow the brain to swell without causing additional damage. Brain injury resulting from primary and secondary effects may result in long-lasting symptoms and clinical manifestations affecting motor, sensory, mood, memory, and higher reasoning functions depending on the locus of the injury.

mTBIs are more difficult to diagnose than moderate and severe cases. A standard magnetic resonance imaging (MRI) scan for an mTBI patient most often appears normal. Thus the diagnosis requires a history of injury and an acute alteration of mental status at the time of injury. Advanced functional neuroimaging techniques, such as diffusion tensor imaging, are now being used to visualize affected areas in patients with mTBI, but these techniques are in early development and are currently used primarily as research tools.

Somatic, cognitive, and emotional/behavioral changes commonly accompany concussion (Figure 2-2), and one or more of these symptoms may be present in a given patient. Post-traumatic stress disorder (PTSD) is a co-morbid condition commonly associated with TBI, and the symptom crossover with TBI can make differentiating the two conditions problematic. Some symptoms, however, are more clearly identified with

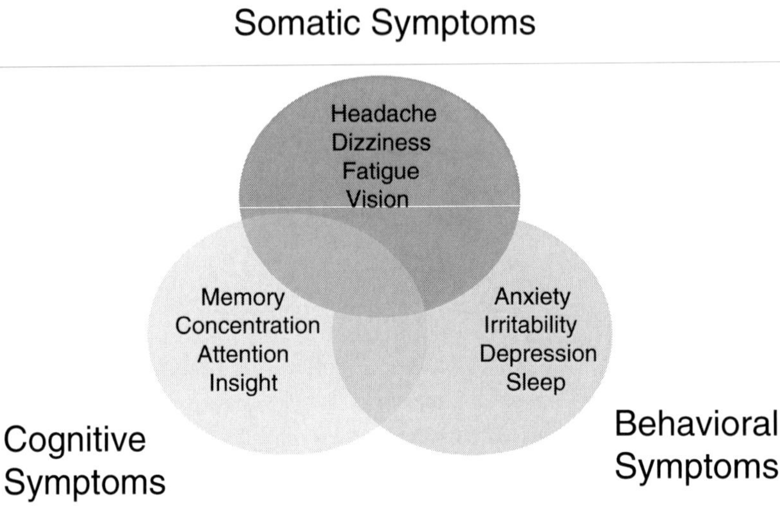

FIGURE 2-2 Categories of post-concussive symptoms. Sources: Arciniegas et al., 2005; Labutta, 2008.

one condition than the other (Figure 2-3). The type and frequency of symptoms reported after an mTBI depends on several factors, especially the severity of TBI and the period of time since the injury. Most individuals with mTBI recover within a few weeks. Studies show that 80 to 90 percent fully recover within one year (Levin et al., 1987), but a small percentage continue to have persistent symptoms (Alexander, 1995).

Post-deployment screening surveys indicate that 10 to 20 percent of military personnel report experiencing an mTBI during their deployment. Most (up to two-thirds) are asymptomatic at the time of the post-deployment survey. The remainder (approximately 5 to 10 percent of the total) acknowledge nonspecific symptoms on their post-deployment survey (TBI Task Force, 2007). It is important to understand that the presence of TBI-related symptoms alone, persistent or not, does not in itself establish a diagnosis of mTBI. The delay in the resolution of these symptoms may be complicated by other factors, such as anxiety, depression, PTSD (Hoge et al., 2008), pain, or the effects of medication.

Proper treatment of mTBI requires early identification, the monitoring and management of symptoms (e.g., headache, dizziness, sleep disturbance), as well as adequate rest prior to return to duty. The latter requirement is important for two reasons. First, reinjury before adequate

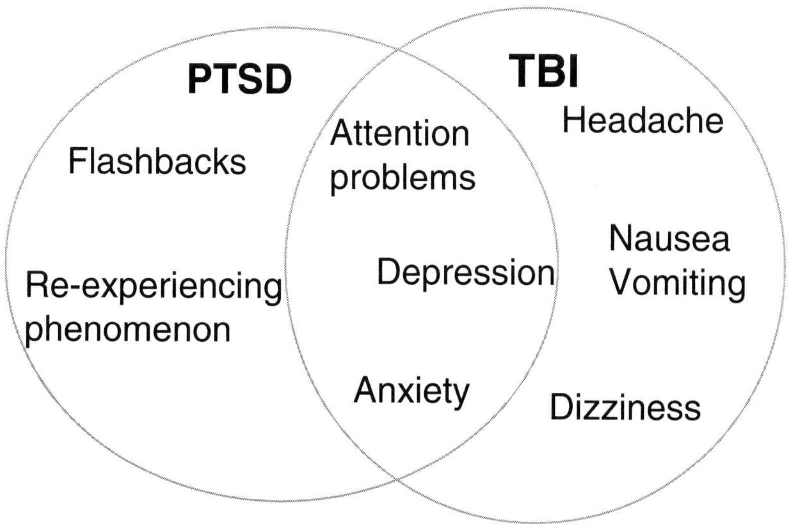

FIGURE 2-3 Distinct and common symptoms of PTSD and TBI. Source: Labutta, 2008.

healing of the initial injury may make the subsequent injury more severe and the recovery of cognitive function slower (Guskiewicz et al., 2003). Second, a soldier or marine whose symptoms affect his or her cognition, such as judgment, concentration, or memory, may jeopardize his or her own safety, as well as the safety and effectiveness of his or her unit.

Moderate and severe cases of TBI are initially treated by controlling secondary factors, such as swelling and bleeding, that can potentially cause further brain damage. For concomitant injuries to the face, neck, or elsewhere, standard surgical or medical treatment is provided to attend to the most critical needs.

Following acute critical care, these TBI patients require both acute and chronic rehabilitative care. Rehabilitation often takes many months and involves a variety of specialty teams, such as physical therapists, occupational therapists, hearing and vision specialists, mental health workers, social workers, and other specialists. The coordination of care by skillful case managers is critical to preventing unnecessary delays and redundant services. In all cases, educating injured service members about their symptoms, instructing them about necessary follow-up, and reinforcing their expectations for full recovery are essential to proper management.

Longer term care often involves moving the patient to another medical facility in the DOD or Department of Veterans Affairs health care system, where the variety of treatments necessary for managing TBI patients can present challenges for both patients and providers. Good coordination and common clinical practice guidelines can improve the quality of care.

Effective family education and support are essential for the smooth, satisfactory management of TBI patients. Family and patient support groups have been shown to be an important ancillary service in promoting recovery. Community-based programs are considered the best approach for long-term management.

The same general principles of medical care apply to TBI patients, whether they are military or civilian. However, care in combat situations must be based on preparing seriously injured patients to be quickly evacuated from the combat zone and preparing less seriously injured patients to return to duty so the military mission can be accomplished. These requirements and the need to employ all able military personnel make the environment of care different for military personnel than for civilians. In the following chapter, the management of TBIs in deployed personnel is described in more detail.

REFERENCES

Alexander, M.P. 1995. Mild traumatic brain injury: pathophysiology, natural history, and clinical management. Neurology 45(7): 1253–1260.

Arciniegas, D.B., C.A. Anderson, J. Topkoff, and T.W. McAllister. 2005. Mild traumatic brain injury: a neuropsychiatric approach to diagnosis, evaluation, and treatment. Neuropsychiatric Disease and Treatment 1(4): 311–327.

CDC (Centers for Disease Control and Prevention). 2006. Traumatic Brain Injury in the United States: Emergency Department Visits, Hospitalizations, and Deaths. Prepared by Division of Injury Response, National Center for Injury Prevention and Control, Centers for Disease Control and Prevention, U.S. Department of Health and Human Services. January 2006. Available online at *http://www.cdc.gov/ncipc/pub-res/TBI_in_US_04/TBI%20in%20the%20US_Jan_2006.pdf* (accessed August 18, 2008).

DOD (U.S. Department of Defense). 2007. Traumatic Brain Injury: Definition and Reporting. Memorandum. HA Policy 07-030. Dated October 1, 2007. Available online at *http://mhs.osd.mil/Content/docs/pdfs/policies/2007/07-030.pdf* (accessed August 13, 2008).

Guskiewicz, K, M. McCrea, S. Marshall, R. Cantu, C. Randolph, W. Barr, J. Onate, and J. Kelly. 2003. Cumulative effects associated with recurrent concussion in collegiate football players: the NCAA Concussion Study. Journal of the American Medical Association 290(19): 2549–2555.

Helmick, K.M., G.W. Parkinson, L.A. Chandler, and D.L. Warden. 2007. Mild traumatic brain injury in wartime. Federal Practitioner 24(10): 58–65.

Hoge, C.W., D. McGurk, J.L. Thomas, A.L. Cox, C.C. Engel, and C.A. Castro. 2008. Mild traumatic brain injury in U.S. soldiers returning from Iraq. New England Journal of Medicine 358(5): 453–463.

Labutta, R.J. 2008. Medical Aspects of Traumatic Brain Injury (TBI). Presentation at the Workshop on Harnessing Operational Systems Engineering to Improve Traumatic Brain Injury Care in the Military Health System, National Academies, Washington, D.C. June 11, 2008.

Langlois, J.A., W. Rutland-Brown, and M.M. Wald. 2006. The epidemiology and impact of traumatic brain injury: a brief overview. Journal of Head Trauma Rehabilitation 21(5): 375–378.

Levin, H.S., S. Mattis, R.M. Ruff, H.M. Eisenberg, L.F. Marshall, K. Tabaddor, W.M. High Jr., and R.F. Frankowski. 1987. Neurobehavioral outcome following minor head injury: a three-center study. Journal of Neurosurgery 66(22): 234–243.

Okie, S. 2005. Traumatic brain injury in the war zone. New England Journal of Medicine 352(20): 2043–2047.

Okie, S. 2006. Reconstructing lives—a tale of two soldiers. New England Journal of Medicine 355(25): 2609–2615.

RAND. 2008. Invisible Wounds of War. Psychological and Cognitive Injuries, Their Consequences, and Services to Assist Recovery, edited by T. Tanielian and L.H. Jaycox. Available online at *http://www.rand.org/pubs/monographs/2008/RAND_MG720.pdf* (accessed August 13, 2008).

Seelig, J.M., D.P. Becker, J.D. Miller, R.F. Greenberg, J.D. Ward, and S.C. Choi. 1981. Traumatic acute subdural hematoma: major mortality reduction in comatose patients treated within four hours. New England Journal of Medicine 304(25): 1511–1518.

Singer, E. 2008. Brain trauma in Iraq. Technology Review 111(3): 52–59.

Warden, D. 2006. Military TBI during the Iraq and Afghanistan wars. Journal of Head Trauma Rehabilitation 21(5): 398–402.

3

Traumatic Brain Injury and the Military Health System[1]

Michael S. Jaffee

Traumatic brain injury (TBI) presents the Military Health System (MHS) with a number of clinical and logistical challenges to treating injuries in the field, transporting injured personnel to higher level care facilities when indicated, and returning injured personnel to active duty or transferring them for therapy in specialized stateside medical centers. Because of the large number of TBI cases in military settings, meeting these challenges is a high priority.

THE MAGNITUDE OF THE CHALLENGE

Data obtained from the Defense and Veterans Brain Injury Center (DVBIC) indicate that nearly a third of all combat injuries treated at the Walter Reed Army Medical Center (WRAMC) in early 2008 involved TBI and that ~20 percent of patients evacuated by air from the Operation Iraqi Freedom (OIF) combat zone were classified as having at least a head or neck injury.

---

[1]This chapter is based on the author's presentation and responses to questions during the plenary session of the NAE-IOM workshop on Harnessing Operational Systems Engineering to Improve Traumatic Brain Injury Care in the Military Health System on June 11, 2008. The author would like to thank IOM staff member David Butler for his assistance in preparing this material for publication.

TABLE 3-1 DVBIC Data—TBI Patient Demographics, by Branch of Service

Branch of Service		Number of Patients	Percentage of Patients
Army	Active duty	3,652	62
	Reserve	203	3
	National Guard	672	11
Marine Corps	Active duty	995	17
	Reserve	73	1
Air Force	Active duty	90	2
	Reserve	2	< 0.5
Navy	Active duty	124	2
	Reserve	25	0.42
	Civilian/NATO	89	2
	Missing data	1	< 0.5
	Total	5,926	

Source: DVBIC data through February 29, 2008.

Many TBIs are the result of blast exposure. Information from the Joint Theater Trauma System (JTTS)[2] indicates that about 40 percent of soldiers exposed to a blast have evidence of a TBI; Defense and Veterans Brain Injury Center (DVBIC) data show that blasts contribute to more than half of combat TBIs.

The majority—some 62 percent—of military TBI patients are members of the active-duty Army, reflecting both their relatively greater numbers in theater and their higher exposure to combat and other hazardous environments. Table 3-1 shows patient demographics by branch of service as of early 2008.

Most diagnosed TBIs (86 percent) are characterized as mild (mTBI); moderate (7 percent), severe (4 percent), and penetrating wounds (3 percent) are far less common. However, these figures may underestimate the true incidence of mTBI. The quantification of moderate, severe, and penetrating TBIs is relatively straightforward because of the frank nature of the injury, but mTBIs are much more difficult to identify. Patients may not immediately experience, recognize, or seek care for mTBI

[2]The Joint Theater Trauma System (JTTS) was established in December of 2004. The implementation of the JTTS involved establishing a theater trauma registry, clinical practice guidelines, and clear lines of communication between the casualties' point of injury through the health care continuum.

injuries, and they or their caregivers (as well as screening mechanisms) may attribute TBI symptoms to other diseases.[3]

To get a better estimate of mTBIs among participants in OIF/ Operation Enduring Freedom (OEF), the military services have been gathering data by various means, such as surveys conducted in theater and immediately after deployment. The VA has also been screening new entrants to its programs, and the RAND Corporation has generated an estimate extrapolated from the results of a telephone survey of previously deployed personnel (RAND, 2008).

All of these sources report that the incidence rate of TBI ranges from 10 to 20 percent (Traumatic Brain Injury Task Force, 2007). The disparities may in part reflect what was being measured and at what point in time it was being measured. Self-reports of TBI symptoms are not the same as positive results from TBI screening instruments or diagnoses of TBI by health professionals. Measurements taken immediately after exposure to potential TBI-initiating events may well generate different figures than measurements taken later, when symptoms may have resolved for some and become manifest for others.

Another confusing element is the extent to which *concussion* is considered a synonym for mTBI. A concussion is defined as a "head trauma-induced alteration in mental status that may or may not involve loss of consciousness" (American Academy of Neurology, 1997). This is different than, although consistent with, DOD's current definition of TBI (DOD, 2007).

Responses by the Department of Defense and Other Federal Agencies

In light of the challenges presented by TBI and other health problems experienced by deployed military personnel, the president, Congress, the federal government in general, and DOD in particular, have all convened high-level groups to identify and address these issues. These include the President's Commission on Care for America's Returning Wounded Warriors, the congressionally chartered Veterans Disability Benefits Commission, the federal interagency Task Force on Returning Global War on Terror Heroes, the DOD Secretary's Independent Review Group on Rehabilitative Care and Administrative Processes at WRAMC

[3]As noted in the preceding chapter, post-traumatic stress disorder and mTBI have some common symptoms.

and National Naval Medical Center, the DOD Office of the Inspector General's Review of DOD/VA Interagency Care Transition, and the Defense Health Board Task Force on Mental Health. A joint DOD/VA Senior Oversight Committee was then formed to "streamline, deconflict, and expedite the two Departments' efforts to improve support of wounded, ill, and injured service members' and veterans' recovery, rehabilitation, and reintegration" (Davis and Day, 2008). Among the lines of action for the Overarching Integrated Product Team, also known as the "Red Cell," was the creation of the Defense Center of Excellence (DCoE) for Psychological Health and Traumatic Brain Injury. DCoE is described as a "Center of Centers," comprising the (aforementioned) DVBIC, National Intrepid Center of Excellence, Center for Deployment Psychology, Deployment Health Clinical Center, and Center for the Study of Traumatic Stress.

Of these, DVBIC—formerly called the Defense and Veterans Head Injury Program or DVHIP—is the longest standing entity, having been established in 1992 as a collaboration between DOD and VA. The original, congressionally directed missions of DVBIC were to provide subject-matter expertise on TBI clinical care, clinical standards, research, and education for DOD and VA. More recently, DVBIC has also assumed responsibility for surveillance of TBI casualties.

As of June 2008, the DVBIC network consisted of 10 military treatment facilities, four VA polytrauma centers, and two civilian partners. Its purpose—and more broadly, a goal of DCoE—is to provide a continuum of high-quality care for all forms of TBI, from the point of injury through acute care, rehabilitation, and chronic care, as well as the transition to disparate providers, as the need arises.

The military's health care delivery for its active force is arrayed along a continuum of five echelons. Table 3-2 lists the echelons and their associated care setting(s) and gives examples of the type of care delivered and the facilities that deliver them.

Echelons of Military Care

The effective triage, stabilization, and transportation of injured persons to the appropriate care facility are vital to their survival. Casualty evacuation (CASEVAC) from the field to initial care in a Level Ib or Level II facility may be performed by whichever land or air vehicle can most quickly deliver the patient to the facility. Tactical (Level II to III) and

TABLE 3-2 Echelons of Care in the Military Health System

Echelon	Care Setting(s)	Care Delivered	Representative Facilities
In Theater			
Level Ia	Field: self aid, buddy aid, combat medic, corpsman	First Aid; airway, hemorrhage control, field dressings, IV fluid, analgesia, stabilization for further evacuation	
Level Ib	Battalion Aid Station		
Level II	Forward Surgical Team (FST); Surgical Shock Trauma Platoon (SSTP), Forward Resuscitative Surgical System	Urgent resuscitative and salvage surgery, stabilization for further evacuation	555th FST, Afghanistan (Patel et al., 2004); SSTP, Al Taqaddum, Iraq (Chambers et al., 2006)
	In-theater clinic (land or ship based)	Minor care, basic laboratory and x-ray services	Sickbay, USS *Abraham Lincoln*
Level III	Combat Support Hospital (CSTH); Air Force Theater Hospital (AFTH); U.S. Navy Hospital Ship	Resuscitation, initial surgery, definitive and reconstructive surgery, post-op care, intensive care, stabilization for further evacuation	325th CSH, Al Asad; 31st CSH, Baghdad; 332nd Expeditionary Medical Dental Group; AFTH, Balad; US Navy Ship *Mercy*
Out of Theater			
Level IV	Regional medical center, general hospital	Definitive care: general and specialized medical and surgical care, reconditioning and rehabilitating services for those returning to duty in theater, stabilization for further evacuation	Landstuhl Regional Medical Center (Germany)
Level V	Medical center	Comprehensive diagnostic, medical, psychological, surgical, and rehabilitative care	Walter Reed Army Medical Center; National Naval Medical Center (Bethesda)

Sources: Jaffee, 2008; Jenkins et al., undated; Rasmussen et al., 2006.

strategic (Level III to IV) medical evacuation (MEDEVAC) is more likely to be carried out by air transports, which have the personnel and equipment to provide en route care and are dedicated to moving patients.

The military's goals are to perform CASEVAC from a (Level I) battalion aid station to Level II care within 38 minutes, tactical evacuation (EVAC) within one hour, and strategic EVAC in 24 to 72 hours. If patients require treatment beyond the level of care available or appropriate at the Level IV Landstuhl Regional Medical Center, they are directed to various sites in the continental United States (CONUS), depending on the type and severity of their injuries. Penetrating TBIs are treated at the National Naval Medical Center; severe and moderate TBIs are sent to Walter Reed or Brooke Army Medical Center; and mild, symptomatic injuries are sent to one of seven regional medical centers. mTBI cases returning to garrison are treated at their duty stations or mobilization sites.

The surgical workload in a Level III care facility—the highest level of care available in theater—is quite different from the workload in a civilian trauma center. The 332nd Expeditionary Medical Dental Group AF Theater Hospital (AFTH) in Balad, Iraq, for example, has approximately four times as many admissions as a typical highest level (Level I) trauma center in the United States. Indeed, virtually all care at Balad AFTH is traumatic care, because the vast majority of patients (> 90 percent) have penetrating traumas (compared to 30 percent in the busiest stateside civilian facility); high-velocity gunshot wounds, blast injuries, and multiple traumas are common. In a stateside civilian facility, less than 10 percent of admitted trauma patients require surgery, and most of those require only one surgical specialty, whereas most patients admitted to the Balad AFTH need surgery, and the majority require more than one specialty and multiple procedures.

COMPONENTS OF AN EFFECTIVE CARE DELIVERY SYSTEM

There are seven core components of an effective program for treating TBI at a military treatment facility:

1. early identification or screening mechanisms
2. an assessment capacity, that is, having enough providers to perform proper assessments

3. a treatment capacity based on proven treatment modalities[4]
4. coordinated care, including a capability for adequate patient follow-up
5. ongoing education for care providers, service members, and patients and their families
6. a link into the DOD TBI surveillance system to enable a better understanding of the scope of the problem
7. a feedback system and tools to maintain and improve unit performance and the quality of care to ensure that the program remains efficient and effective over time

Several factors can complicate the delivery of TBI care at military facilities. For example, TBI requires a multidisciplinary clinical response. Unlike heart disease and cardiology, no single specialty has the default responsibility for the treatment of TBI. Depending on the specific injuries, several departments, including neurology, neurosurgery, ophthalmology, otolaryngology, physical medicine and rehabilitation, psychiatry, and psychology, as well as multiple therapy units, such as physical, occupational, and speech therapy, could plausibly take the lead. As a result, there is little consistency in care management among and between facilities.

A second factor is the relatively high rate of personnel turnover as the result of tour rotations, [re]deployments, and separations at the end of service. Thus it is difficult to train and retain skilled professionals at particular care locations.

Capacity, a third factor, noted in the discussion of DVBIC above, is being addressed through partnerships with VA and civilian entities. The challenge of such partnerships is that non-DOD facilities may not apply the same standards of care or may not have the same level of expertise as military facilities. Accreditation, through, for example, the Commission for Accreditation of Rehabilitation Facilities, is one way to manage this issue.

The educational component of a program for TBI includes clinical training for providers who render care in theater, during pre-deployment and deployment, and back in CONUS. The training includes programs intended to increase the awareness of TBI by service members and their commanders, not only to improve care, but also to reduce the stigma

---

[4]In general, the clinical information (Class 1 randomized controlled trial) on treating severe TBI is better than the information on treating mTBI.

associated with TBI injuries. Providing patients and their families with information about TBI prognosis and recovery expectations and DOD services helps them cope with the injury. The educational mission can be extended to reserve and guard components of the military and to the civilian community.

Care coordination is another vital component of TBI medical services. Good coordination can ensure that patients do not "fall through the cracks" as they make their way through different levels of care and between facilities. Both DOD and VA have established systems for coordinating care since 2006. Expert care coordination for TBI patients, one of DVBIC's goals,[5] is delivered through a network of regional care-coordination sites, 14 of which were located in the United States and Germany as of early 2008. More sites are planned to increase their geographic focus in areas where demand is high.

VA's care-coordination system is intended to facilitate services for patients with polytraumatic injuries in general, with special emphasis on TBI. There are four levels of care in the VA system (which should not be confused with DOD's levels of care). The highest, Level I, is provided at four polytrauma rehabilitation centers (PRCs) located in Minneapolis; Palo Alto; Richmond, Virginia; and Tampa.[6] PRCs include brain injury centers (VA's component of DVBIC), which provide a full range of acute, comprehensive medical and rehabilitative services to patients with highly complex injuries, including an emerging consciousness program for patients with severe TBI to facilitate their return to awareness and improve responsiveness.

PRCs are augmented by 21 Level II polytrauma network sites (one in each of the VA's Veterans Integrated Service Network regions) that provide inpatient and outpatient post-acute rehabilitation in settings closer to veterans' homes (VA, 2008). The 76 Level III polytrauma support clinic teams (PSCTs) are composed of rehabilitation providers who supply long-term polytrauma management, primarily on an outpatient basis. PSCTs also head the comprehensive TBI screening, performing a four-question preliminary screen (with appropriate follow-up) for every OIF/OEF veteran who requests any sort of care. As of fall 2007, 54 Level IV polytrauma points of contact (PPOC) were available at VA facilities that do not provide higher level care. These PPOCs are

---

[5] *http://www.biausa.org/elements/pdfs/awareness/dvbic_fact_sheet.pdf.*

[6] In 2007, Congress directed that a fifth center be established in San Antonio. See *http://www.usmedicine.com/article.cfm?articleID=1658&issueID=104.*

intended to serve as local sources of expertise on the entire polytrauma system of care and provide referrals to appropriate providers and facilities (Sigford, 2007).

Data on patients in the MHS formerly were collected through disparate, service-specific systems. The JTTS, which is under the Office of the Secretary of Defense, consolidated these into a single Joint Theater Trauma Registry. The objective of the registry is to log information at a patient's first point of contact with MHS and follow that person forward. Modules are being developed for the system to gather specific information on some outcomes, including treatment for TBI.

JTTS staff positioned at each of the Level III hospitals in theater and some of the major CONUS hospitals collect data. However, the system is in its infancy and—as of spring 2008—there was a backlog of some 17,000 charts to be entered into the database. These data can be important because symptoms may have clinical significance long after they are reported or have seemingly been resolved. The Medical Communications for Combat Casualty Care system and its Armed Forces Health Longitudinal Technology Application-Theater are also intended to systematize the collection of medical information in the field. JTTS also facilitates other DOD-wide collaborations in trauma management, policy development, research, education, medical resource allocation, and clinical care (DOD, 2008).

Both DOD and VA are advancing the use of telehealth technologies to help deliver services to patients in theater and in rural, nonurban, and underserved areas in the United States. Two current applications are teleconsultations that connect experts on TBI to on-site care providers and teleconferencing that enables patients and their families to meet their VA treatment team before they transition out of DOD care. Virtual support groups, which will link patients in disparate locations, are in the planning stage.

RESEARCH QUESTIONS AND INITIATIVES

The TBI clinical-care delivery system is supported and enriched by a framework of scientific and medical research by agencies throughout DOD. This framework has three primary elements that track the timeline from exposure to potentially hazardous conditions to the decision to return an injured warrior to duty or separate him or her from service.

The three elements are: (1) protection and prevention, (2) clinical assessment, and (3) injury management.

Protection and Prevention

One component of the protection and prevention element is helmet design. In the past, the driving consideration was the prevention or mitigation of a penetrating injury. This is, of course, still vital, but the considerations have been expanded to include the mitigation of blast effects. Sensors placed in a helmet can provide information about its acceleration, a proxy for acceleration of the head and brain during a concussive event. The goal of this research is to develop a blast dosimeter that can provide an objective measure of the force experienced and yield information about how injuries vary with the magnitude of the force. Lessons are also being learned from how well materials used in athletic helmets protect against blunt-force trauma.

A more general research question is the mechanism of blast injury, which can result from the detonation of a vehicle-borne or person-borne explosive device, rocket-propelled grenade (RPG), or improvised explosive device (IED) (DOD, 2008). Relatively little is known about such injuries, both because explosives were not used as primary weapons in conflicts prior to OIF/OEF and because modern body armor now saves the lives of many more people exposed to blasts.

Table 3-3 categorizes blast injuries by mechanism. Briefly, a primary blast injury is caused solely by the direct effect of blast overpressure on tissue. Because air is easily compressible, a primary blast injury almost always affects air-filled structures, such as the lung, ear, and gastrointestinal tract. Secondary blast injuries are caused by objects propelled by the force of the explosion. These include fragments of shell casings and materials in the surrounding environment that are thrown or fragmented by the blast. The impact of these objects can cause closed and open TBIs.[7]

Tertiary injuries, a feature of high-energy explosions, occur when the force of a blast propels a person into objects in the surrounding environment. DOD includes skin speckling from the residue of explosive products and blunt and crush injuries caused by blast-collapsed structures in this category (DOD, 2006). Quaternary effects result from

[7]An "open-head" TBI is caused by a penetrating wound; the skull remains intact in a "closed-head" injury.

TABLE 3-3 Blast Injuries by Mechanism

Category	Mechanism	Representative Effects
Primary	Direct exposure to blast overpressure	• Eardrum rupture and middle ear damage • Blast lung injury (BLI)[a] • Abdominal hemorrhage and perforation • Eye rupture
Secondary	Impact of blast-energized debris from the device and surrounding environment	• Penetrating ballistic (fragmentation) or blunt injuries • Eye penetration
Tertiary	Displacement of the person by the blast or debris impact	• Fracture and traumatic amputation • Stripping of soft tissues • Skin speckling with explosive product residue • Blunt injuries • Crush injuries
Quaternary	Exposure to the heat and fire by-products generated by the blast	• Burns from radiant and convective heat • Injury or incapacitation from inhaled toxic fire gases
Quinary[b]	Exposure to toxic agents released by the blast	• Illnesses, injuries, or diseases caused by chemical, biological, or radiological substances (i.e., "dirty bombs")

[a]"Blast lung injury (BLI) . . . is a direct consequence of the blast wave from high explosive detonations upon the body. . . . The blast wave's impact upon the lung results in tearing, hemorrhage, contusion, and edema with resultant ventilation-perfusion mismatch" (CDC, 2008).
[b]Note that some sources (CDC, 2006) combine quaternary and quinary injuries into an "all other" category, and others define the quinary mechanism and effects differently (Kluger et al., 2007).
Sources: DOD, 2006; Jaffee, 2008; Warden, 2006.

exposure to the heat and chemicals generated by a blast, that is, explosive by-products. Quinary effects are caused by nuclear, biological, or chemical agents released by the blast, either because they were included with the charge or because they were in the environment around it.

The means by which a blast directly induces a closed-head injury is a subject of scientific debate. Among the theories that have been advanced are the formation of air emboli in blood vessels, small-scale cavitation,[8] exposure to blast-generated electromagnetic fields, and overpressure or underpressure effects.

The last of these has received a great deal of attention. Overpressure and underpressure waves[9] can cause ruptures of the tympanic membranes in the ears, a mechanism called barotrauma. There are two hypotheses for how these waves are transmitted to the brain: directly through the skull, eyes, and ears; and indirectly through the great vessels of the chest (Bhattacharjee [2008] reporting on the research of Ibolja Cernak et al., Johns Hopkins University).

Empirical modeling of blast effects has yielded estimates of injury types as a function of distance from the event. However, the value of these estimates depends on how representative they are of real-world conditions. Open-air blasts, for example, are the easiest to model, but battlefield environments are often cluttered with buildings and vehicles that can channel, reflect, and redirect explosive forces.

Animal models are also being used to model blast effects, using shock tubes to transmit the energy of detonations and deflagrations.[10] These studies have shown axonal damage, swelling, edema, and reactive gliosis. Rafols and colleagues (2004) have reported a genetic change, alternation in the expression of inducible nitric oxide synthetase.

Another complication in the analysis of TBI in combat situations is that most blast injuries involve other modalities, a circumstance that has been called "blast plus." A soldier riding in a HMMWV (popularly

---

[8]Cavitation is the sudden formation and collapse of bubbles in liquids (blood or cerebral fluid, for example) caused by a mechanical force such as violent shaking.

[9]A blast generates a shock front that compresses the air, creating an initial overpressure wave. This is followed by a negative (underpressure) phase that results from the vacuum behind the shock front. A second, smaller overpressure may in turn be generated by the underpressure, and so on as the wave dissipates. These wave phases may cause different types of damage to the brain.

[10]A *detonation* induces a shock wave that propagates at supersonic speed, whereas a *deflagration* generates a subsonic wave. Detonations thus create higher pressures and are more destructive.

referred to as a "Humvee"), for example, may experience a blast that throws shrapnel and causes his vehicle to suddenly decelerate, whipping the head of its occupants and possibly driving them into the roof, all of which may contribute to brain injuries. In contrast, TBIs in other circumstances, such as motor vehicle accidents or sports collisions, are typically single-mechanism injuries. For this reason, the extent to which lessons learned in civilian cases are applicable to the military, including whether the pathophysiology of injury, pattern of co-morbidities, and natural history of recovery, are debatable.

Clinical Assessment

Military health care providers also face many other controversies. Perhaps the most basic is the definition of TBI itself. The definition of mTBI currently used by DOD and VA is based on a definition developed by the American Congress of Rehabilitation Medicine that identifies a person as having mTBI if he or she experiences either a loss of consciousness or an alteration of consciousness (GAO, 2008).[11] Disagreements about whether to use only loss of consciousness or either loss *or* alteration of consciousness have not been resolved.

Research indicates that "[i]njuries associated with loss of consciousness carried a much greater risk of health problems than [did] injuries associated with altered mental status" (Hoge et al., 2008). However, there is wide consensus that altered consciousness ought to be a diagnostic criterion (DVBIC Working Group, 2006),[12] and nearly two-thirds of military mTBI cases qualify for the diagnosis on the basis of alteration of consciousness. Thus, if this criterion is dropped, it may lead to missed cases and the negative health consequences that flow from them. The differential diagnosis of mTBI and the identification of co-morbidities pose additional challenges.

---

[11]Specifically, the standards define a person as having mTBI if "the person had a traumatically-induced physiological disruption of brain function as demonstrated after an event by at least one of the following: (1) any period of loss of consciousness; (2) any loss of memory for events immediately before or after the event; (3) any alteration in mental state at the time of the event, for example feeling dazed, disoriented, or confused; and (4) a focal neurological deficit or deficits that may or may not have been transient, for example loss of coordination, speech difficulties, or double vision" (GAO, 2008, p. 9).

[12]U.S. Central Command subsequently mandated the use of clinical guidelines that included the alteration of consciousness criterion (Casscells, 2008).

Another controversy concerns the utility of post-deployment screenings, which are performed as part of the Post-Deployment Health Assessment and Re-Assessment program. Although some studies exist (Schneiderman et al., 2008; Schwab et al., 2007), screening instruments used by the military have not, by and large, been validated in military populations. DVBIC is conducting outreach to military personnel, their families, and medical service providers to educate them about mTBI signs and symptoms and thus complement screening efforts (RAND, 2008).

Several panels and commissions, including the Armed Forces Epidemiological Board[13] (AFEB, 2006), the Army Surgeon General's TBI Task Force (Traumatic Brain Injury Task Force, 2007), a group constituted by DOD to review operations at Walter Reed and National Naval Medical Centers (Independent Review Group, 2007), and the Defense Health Board, an advisory panel convened under the Office of the Assistant Secretary of Defense for Health Affairs (OSD/HA) (DOD Task Force on Mental Health, 2007), have advised DOD either to consider or implement a combination of baseline and post-deployment neurocognitive screening, or post-injury testing. In addition, the National Defense Authorization Act for Fiscal Year 2008 (P.L. 110-181) directed DOD to develop and deploy an "evidence-based means of assessing traumatic brain injury . . . including a system of pre-deployment and post-deployment screenings of cognitive ability in members for the detection of cognitive impairment" (§1618).

All of these recommendations and directives have influenced the system of care. Current policy requires pre-deployment baseline screening using the Automated Neuropsychological Assessment Metric, an Army-developed instrument for which normative data from military populations are available. Other instruments are also being considered by DOD.

Separately, DOD is evaluating information technologies and systems issues related to testing. The Army, Navy, and Air Force use different software and hardware, making it difficult to implement a single automated system for all of the services and for software written for one branch of the military to communicate or be combined with software written for another.

[13]The AFEB is now a part of the Defense Health Board, which also comprises the former Amputee Patient Care Program Board of Governors and the Armed Forces Institute of Pathology Scientific Advisory Board.

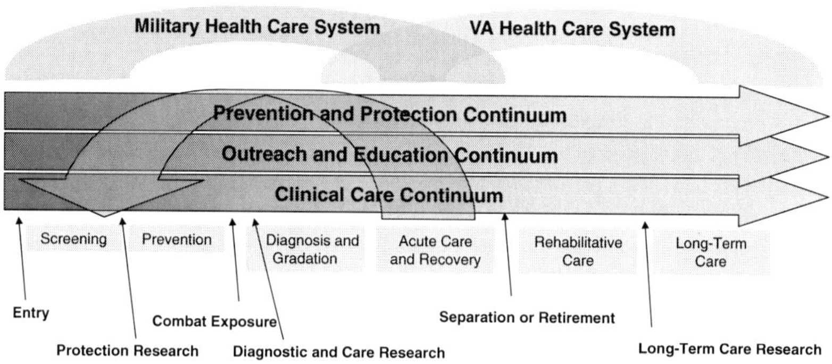

FIGURE 3-1 TBI Health Care Delivery and Research Collaboration System. Source: Jaffee, 2008.

Injury Management

The integration of care delivery within and among the services and between them and other providers is an important goal for TBI management. DOD is taking steps to encourage systems collaboration and facilitate the interface between MHS and VA health care systems. Figure 3-1 shows the major components of the care and research network. Systems engineering technologies and techniques have the potential to make important contributions in this vital area.

REFERENCES

AFEB (Armed Forces Epidemiology Board). 2006. Traumatic Brain Injury in Military Service Members–2006-02. Memorandum for the Honorable William Winkenwerder, Jr., M.D., Assistant Secretary of Defense for Health Affairs. Available online at *http://www.ha.osd.mil/afeb/2006/2006-02.pdf* (accessed August 11, 2008).

American Academy of Neurology. 1997. Practice parameter: the management of concussion in sports. Neurology 48(2): 581–585.

Bhattacharjee, Y. 2008. Shell shock revisited: solving the puzzle of blast trauma. Science 319(5862): 406–408.

Casscells, S.W. 2008. Statement on mental health by the Honorable S. Ward Casscells, M.D., Assistant Secretary of Defense for Health Affairs before the Subcommittee on Military Personnel, Armed Services Committee, U.S. House of Representatives, March 14, 2008. Available online at *http://armedservices.house.gov/pdfs/MilPers031408/Casscells_Testimony031408.pdf* (accessed October 22, 2008).

CDC (Centers for Disease Control and Prevention). 2006. Explosions and Blast Injuries: A Primer for Clinicians. Available online at *http://www.bt.cdc.gov/masscasualties/explosions.asp* (accessed July 30, 2008).

CDC. 2008. Blast Injuries: Blast Lung Injury. Available online at *http://www.bt.cdc.gov/masscasualties/blastlunginjury.asp* (accessed July 30, 2008).

Chambers, L.W., D.J. Green, B.L. Gillingham, K. Sample, P. Rhee, C. Brown, S. Brethauer, T. Nelson, N. Narine, B. Baker, and H.R. Bohman. 2006. The experience of the US Marine Corps' Surgical Shock Trauma Platoon with 417 operative combat casualties during a 12 month period of Operation Iraqi Freedom. Journal of Trauma-Injury Infection and Critical Care 60(6): 1155–1161; discussion 1161–1164.

Davis, L.C., and K. Day. 2008. Statement of Dr. Lynda C. Davis, Deputy Assistant Secretary of the Navy for Military Personnel Policy, Department of Defense, and Ms. Kristin Day, Chief Consultant, Care Management and Social Work, Department of Veterans' Affairs, before the U.S. Senate Committee on Veterans' Affairs, March 11, 2008. Available online at *http://veterans.senate.gov/public/index.cfm?pageid=16&release_id=11536&sub_release_id=11593&view=all* (accessed August 13, 2008).

DOD (U.S. Department of Defense). 2006. Department of Defense Directive Number 6025.21E: Medical Research for Prevention, Mitigation, and Treatment of Blast Injuries. July 5, 2006. Available online at *http://www.dtic.mil/whs/directives/corres/pdf/602521p.pdf* (accessed July 30, 2008).

DOD. 2007. Traumatic Brain Injury: Definition and Reporting. Memorandum. HA Policy 07-030. Dated October 1, 2007. Available online at *http://mhs.osd.mil/Content/docs/pdfs/policies/2007/07-030.pdf* (accessed August 13, 2008).

DOD. 2008. Operational Medicine and Medical Force Readiness (OM&MFR). Available online at *http://deploymentlink.osd.mil/about.jsp?topic=6#ttm* (accessed July 22, 2008).

DOD Task Force on Mental Health. 2007. An Achievable Vision: Report of the DOD Task Force on Mental Health. Falls Church, VA: Defense Health Board.

DVBIC Working Group (Defense and Veterans Brain Injury Center Working Group on the Acute Management of Mild Traumatic Brain Injury in Military Operational Settings). 2006. Clinical Practice Guideline and Recommendations 22 December 06. Available online at *http://dvbic.org/public_html/pdfs/clinical_practice_guideline_recommendations.pdf* (accessed August 7, 2008).

GAO (Government Accountability Office). 2008. VA Health Care: Mild Traumatic Brain Injury Screening and Evaluation Implemented for OEF/OIF Veterans, but Challenges Remain. GAO-08-276. Available online at *http://www.gao.gov/new.items/d08276.pdf* (accessed September 10, 2008).

Hoge, C.W., D. McGurk, J.L. Thomas, A.L. Cox, C.C. Engel, and C.A. Castro. 2008. Mild traumatic brain injury in U.S. soldiers returning from Iraq. New England Journal of Medicine 358(5): 453-463.

Independent Review Group on Rehabilitative Care and Administrative Processes at Walter Reed Army Medical Center and the National Naval Medical Center. 2007. Rebuilding the Trust. Available online at *http://www.washingtonpost.com/wp-srv/nation/documents/walter-reed/IRG-Report-Final.pdf* (accessed August 11, 2008).

Jaffee, M.S. 2008. TBI Overview in Military Health System. Presentation at the Workshop on Harnessing Operational Systems Engineering to Improve Traumatic Brain Injury Care in the Military Health System, The National Academies, Washington, D.C., June 11, 2008.

Jenkins, D.H., R.M. Bolenbaucher, and T. Cortner-Pouncy. n.d. Trauma System Development in a Combat Theater. Available online at *http://www.amtrauma.org/courses/article_combatTheater.doc* (accessed July 18, 2008).

Kluger, Y. A. Nimrod, P. Biderman, A. Mayo, and P. Sorkin. 2007. The quinary pattern of blast injury. American Journal of Disaster Medicine 2(1): 21-25.

Patel, T.H., K.A. Wenner, S.A. Price, M.A. Weber, A. Leveridge, and S.J. McAtee. 2004. A U.S. Army Forward Surgical Team's experience in Operation Iraqi Freedom. Journal of Trauma-Injury Infection and Critical Care 57(2): 201–207.

Rafols, D., J. Steiner, J.A. Rafols, and T. Petrov. 2004. Intracellular coexpression of endothelin-1 and inducible nitric oxide synthase underlies hypoperfusion after traumatic brain injury in the rat. Neuroscience Letters 362(2): 154–157.

RAND. 2008. Invisible Wounds of War. Psychological and Cognitive Injuries, Their Consequences, and Services to Assist Recovery, edited by T. Tanielian and L.H. Jaycox. Available online at *http://www.rand.org/pubs/monographs/2008/RAND_MG720.pdf* (accessed August 13, 2008).

Rasmussen, T.E., W.D. Clouse, D.H. Jenkins, M.A. Peck, J.L. Eliason, and D.L. Smith. 2006. Echelons of care and the management of wartime vascular injury: a report from the 332nd EMDG/Air Force Theater Hospital, Balad Air Base, Iraq. Perspectives in Vascular Surgery and Endovascular Therapy 18(2): 91–99.

Schneiderman, A.I., E.R. Braver, and H.K. Kang. 2008. Understanding sequelae of injury mechanisms and mild traumatic brain injury incurred during the conflicts in Iraq and Afghanistan: persistent postconcussive symptoms and posttraumatic stress disorder. American Journal of Epidemiology 167(12): 1446–1452.

Schwab, K.A., B. Ivins, G. Cramer, W. Johnson, M. Sluss-Tiller, K. Kiley, W. Lux, and B. Warden. 2007. Screening for traumatic brain injury in troops returning from deployment in Afghanistan and Iraq: initial investigation of the usefulness of a short screening tool for traumatic brain injury. Journal of Head Trauma Rehabilitation 22(6): 377–389.

Sigford, B. 2007. VHA Polytrauma / TBI System of Care. Presentation before the American Academy of Physical Medicine and Rehabilitation, September 29, 2007. Available online at *www.aapmr.org/zdocs/member/warriors/warriors_VHA.ppt* (accessed July 29, 2008).

Traumatic Brain Injury Task Force. 2007. Report to the (Army) Surgeon General: Traumatic Brain Injury Task Force, May 15, 2007. Available online at *http://www.armymedicine.army.mil/reports/tbi/TBITaskForceReportJanuary2008.pdf* (accessed September 10, 2008).

VA (Department of Veterans Affairs). 2008. VA Polytrauma System of Care: Facility Locations. Available online at *http://www.polytrauma.va.gov/facility_locations.asp?isFlash=1* (accessed July 29, 2008).

Warden, D. 2006. TBI—Current Trends, Strategies to Diagnose, Treat, and Rehabilitate Casualties. Presentation to the Institute of Medicine Board on Military and Veterans Health, December 18, 2006.

4

Examples of Operational Systems Engineering Applications Relevant to Traumatic Brain Injury Care[1]

William P. Pierskalla

Operational systems engineering (OSE), which combines science and mathematics to improve the operations of systems and enterprises that provide goods and services, entails describing, analyzing, planning, designing, and integrating complex systems. Operational systems engineers use deterministic and probabilistic mathematics (called stochastic processes) to describe how systems operate, design systems based on those descriptions, and integrate all elements of systems operations, including people, processes, materials, equipment, and facilities, to improve efficiency and effectiveness.

The tools, techniques, and concepts of OSE have been applied to many different areas of health care (NAE/IOM, 2005):

- Epidemiology, health promotion, disease prevention, and predictions of the incidence, prevalence, and mortality of diseases. OSE has been used to evaluate intervention strategies, disease-control programs, and screening programs.
- Health-care and health-systems design, including estimates of future resource needs and the deployment of those resources.

[1]This chapter is based on the author's presentation and responses to questions during the plenary session of the NAE-IOM workshop on Harnessing Operational Systems Engineering to Improve Traumatic Brain Injury Care in the Military Health System on June 11, 2008. The author would like to thank NAE staff member Proctor Reid for his assistance in preparing this material for publication.

OSE has been used to design service systems (e.g., kinds of services offered, technologies, site/location selection); optimize capacity planning for health-delivery facilities; select approaches to business/enterprise planning; and design emergency services.
- Support of medical decision making. OSE has been used for prevention programs, modeling of diseases, optimizing diagnostic tools, optimizing therapy programs and chronic care programs, and genetic modeling.
- Operation of health care systems. OSE has been used to develop and implement appointment systems and waiting times for outpatients/inpatients; determining staff levels and scheduling; conducting inventories, material requirements planning, supply chains; forecasting short- and long-term demand; planning auxiliary services; and evaluating medical technologies.

In this chapter, we review five examples of OSE modeling used to address challenges to health care and health care systems relevant to the major issues in traumatic brain injury (TBI) care and management. Four of these examples were presented and discussed during the workshop; the fifth, which was mentioned briefly in the plenary discussion, is included because of its direct relevance to TBI care.

The first example is based on research by Hazen (2004) on using dynamic influence diagrams in medical decision making. The term "dynamic influence diagram" applies to a process with a stochastic (i.e., random or probabilistic) component. Bayesian networks, a subset of dynamic influence diagrams, are a means of graphically representing the relationship between a set of variables and their varying probabilistic interdependencies.

The second example involves strategies for screening blood for the presence of the human immunodeficiency virus (HIV) antibody (Schwartz et al., 1990). This example shows how OSE can address the conditional sensitivities and specificities of testing, which are directly applicable to diagnosing and treating patients with mild TBI (mTBI).

The third example illustrates an approach to policy decision modeling of the costs and outputs of medical-school education, a large complex problem with both economic and policy ramifications, as well as strategic and operational issues (Lee et al., 1987).

The fourth example illustrates the large-scale application of an OSE simulation to a geographically dispersed health care delivery enterprise

(Bonder, 2005). The model was built for the Military Health System (MHS) during the late 1990s to address a large number of issues related to capacity, organization, resource allocation, and process change.

The final example illustrates the direct application of OSE modeling to the management of TBI patients in the Department of Veterans Affairs (VA) medical system.

EXAMPLE 1
DYNAMIC INFLUENCE DIAGRAMS FOR MEDICAL DECISION MAKING

Influence diagrams, which have been used for modeling in decision analysis for some time, facilitate the analysis of sequential decisions. In Hazen's paper, "Dynamic Influence Diagrams: Applications to Medical Decision Making," he uses dynamic influence diagrams to structure and analyze a continuous chain of decisions related to whether or not a patient should proceed with total hip replacement surgery in a context in which back-stepping loops are possible (Hazen, 2004). In other words, once a decision is made, things might happen that require revisiting that decision and then moving forward again. Thus a dynamic influence diagram models a looping, continuous, recycling decision process (Figure 4-1).

In this example, Hazen was dealing with a simple decision between two choices—whether a patient should elect to have a total hip arthroplasty (THA) (i.e., total hip replacement) or should opt for conservative management (i.e., no medical intervention). The purpose of the model is to calculate the optimal expected quality-adjusted lifetime for both choices. Although many variables can be included in the model, such as age, sex, mobility, and/or other functional, social, demographic, and racial characteristics of the patient, the main variables in this example are race, age, and sex.

The operation of the model begins when the patient is given a diagnosis of American College of Rheumatology (ACR) Class III osteoarthritis. At that point, a decision must be made by patient and doctor as to the appropriate therapy. Following the therapy, sooner or later, the patient will either transition to one of the other three ACR classes, will generally deteriorate, or will die (Box 4-1).

The transition to another ACR class or to death is modeled based on certain probabilities, depending on the characteristics of the patient

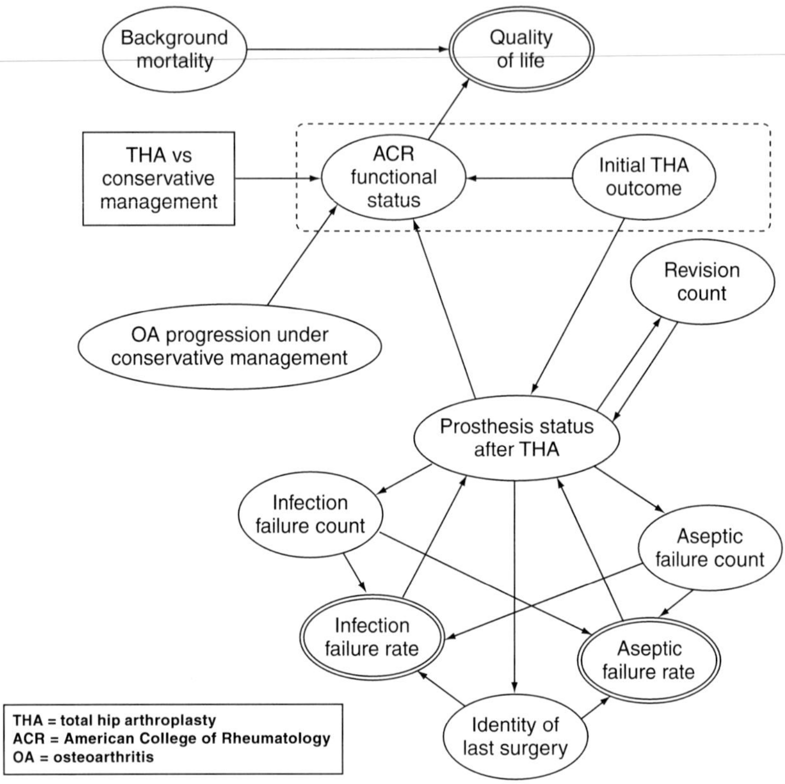

FIGURE 4-1 A dynamic influence diagram for a model of the choice between THA and conservative management. Source: Adapted from Hazen, 2004.

and, possibly, the characteristics of the surgical procedure, such as the hospital where it was performed and the experience of the surgeon. Although the characteristics of the surgery are not shown in this model, they could easily be included.

After being reclassified in the new ACR class, the patient may remain in that class for some time and then, later, transition to another class via infective or aseptic failure of the hip prosthesis or to death from another cause or to a general deterioration of his or her health.

Considering all of these possibilities, which can happen randomly over time with certain probabilities, the prognosis of the patient's future quality of life is determined by a reverse analysis from the last point in the patient's life to the beginning point where the decision must be made as to surgery or conservative management. Also using reverse analysis,

BOX 4-1
American College of Rheumatology
Functional Classifications for Hip Osteoarthritis

Class	Description
I	Complete ability to carry on all usual duties without handicap
II	Adequate for normal activities despite handicap of discomfort or limited motion in the hip
III	Limited to little or none of duties of usual occupation or self-care
IV	Incapacitated, largely or wholly bedridden or confined to wheelchair, little or no self-care

Source: Hazen, 2004.

the total lifetime costs of THA and conservative management can be calculated and minimized. Therefore, a patient with given characteristics has an expected quality of life depending on the decision made at the beginning point. Using the model to work back to that point, the best (optimal) decision can be chosen to maximize the remaining quality of life.

If other types of therapy are available or become available in the future, they can be included in the model and analyzed as to their outcomes for expected quality of life. For example, cortisone shots or cartilage replacement might be considered in the model.

In Hazen's example, he used Cox stochastic trees to predict what the Cox progression to death and the stages would be for an 85-year-old white male and for a 60-year-old white female. A stochastic tree is a graphical modeling tool that allows the explicit factoring of temporal uncertainty—for example, age-dependent mortality rates—in a decision tree analysis.

An 85-year-old white male has to choose one of two options—THA or conservative management. Using the model, if he chooses THA, his quality of life would be in ACR Class I for about 2.9 years on average. If Class II were the result of the THA, he would have 1.4 years in this class. If Class III were the result, he would have about

.06 years in this class. If he chose conservative management, he could expect to be in Class III for 3.9 years and then proceed to Class IV until death.

THA would not give this patient more years of expected life than conservative management because of his advanced age—his life expectancy is about 4.4 years with either decision. However, the patient's mean quality of life, when adjusted for Class I for quite a long time, Class II for a fair amount of time, and so on, would be significantly better than if he opted for conservative management.

Hazen estimated that THA at the time this research was conducted (in the mid-1990s) would cost the 85-year-old $25,000 in fixed costs plus another $5,000 for costs associated with possible revision surgery, for a total of $30,000. With conservative management, he would spend $20,000 for home and/or nursing care prior to dying. By these criteria, for an additional $10,000, the patient would buy a better quality of life for about $5,000 per additional quality-adjusted life year (QALY), far less than the amount a person would spend for additional QALYs with a coronary artery bypass or a stent implant. So, the model shows that for this person, at age 85, THA at $5,000 per QALY is a good deal.

Applying the same analysis for a 60-year-old white female considering THA, her mean quality of life would be 9 years in Class I and 5 years in Class II (a total of 14 years of high quality) as opposed to 7 years in Class III. The total cost for THA would be about $54,000. However, if the patient decided to forego the surgery, the estimated cost, taking into account the attendant problems with her hip, medication, and long-term care, would be about $174,000 for care over her lifetime. In this case, she would actually save money by undergoing a THA.

Relevance to TBI Care Management

For an mTBI patient, decisions for therapies and the timing of treatments might be considered as a flow of the patient through different mTBI states. The transitions over time (considered probabilistically) could be based on repeated traumas in the field and by the characteristics of the treatment provided.

Mild TBI, however, is more complex than THA; each injury is different and results in different physical, psychological, and functional effects. The additional complexity can be indicated in the model with more branching nodes and arcs. Decisions and outcomes are expressed stochastically. Patients and providers must then make continuing decisions about when, how, and what to do next.

The model could begin with the field incident, the decisions made at that time, subsequent transfers and decisions, subsequent discharge and decisions, and so forth over time. Or the model could begin with a later discovery of mTBI and the flows of decisions, therapies, and outcomes (again probabilistically) from that point on. The model offers many combinations of timing and choices of therapies, locations, personnel, and modes of care delivery.

It is also possible to use different or multiple criteria to model the decisions that would optimize a patient's predicted quality of life. In the THA example, the criteria were (1) to maximize the quality of lifetime remaining and (2) to minimize the costs to the individual and/or society. Similar or different criteria could be used to optimize mTBI treatment plans, locations, and/or personnel.

EXAMPLE 2
SCREENING BLOOD FOR THE HUMAN IMMUNODEFICIENCY VIRUS ANTIBODY

The development and application of a decision analytic model to examine alternative strategies for screening blood for the HIV antibody and making decisions affecting blood donors was presented as having strong parallels to testing for mTBI and making decisions about next steps for wounded soldiers (Schwartz et al., 1990).

At the time this work was done, in the mid 1980s, limited knowledge was available about the biology, epidemiology, and early blood manifestations of HIV. Furthermore, the initial and conditional sensitivities and specificities of enzyme immunoassays (EIA) and western blot (WB) tests had wide ranges of error. Finally, nothing was then known of the effectiveness of registries, counseling of donors, self-reporting of donors' sexual and drug-injection activities, or related educational programs.

The purpose of the model was to determine which screening tests to use, and in what sequence, to minimize the number of HIV-infected units of blood and blood products entering the nation's blood supply at an acceptable cost. The model is a Bayesian decision-tree model, and the decisions are probabilistically based.

The variables in the model are: available screening tests, the initial and conditional ranges of sensitivities and specificities of the tests at different levels of HIV (based on time since incidence), the costs of administering tests, the costs of informing donors of positive results, the costs of maintaining a donor registry of positives, and the prevalence and incidence of HIV in different donor populations (Figure 4-2).

Given a unit of blood to test, a decision must be made as to which screening test to use initially (an EIA or the WB test). Based on the results of the first test, a decision must then be made to accept the unit into the blood supply (test negative) or to conduct another test (initial test positive) to determine (still probabilistically) if the positive result is correct. Following the second test, subsequent decisions are made: conduct a third test; accept the unit into the blood supply; reject the unit but do not inform the donor; reject the unit and inform the donor; enter the donor in a registry of individuals whose blood will not be accepted into the blood supply in the future.

Although more than three tests might be conducted, the first two provided enough information to make informed decisions based on the conditional probabilities of the presence or absence of HIV. The model provides the following outputs:

- the expected number of infected units entering the blood supply during a specific period of time (e.g., a year)
- the expected number of uninfected units (good) discarded (i.e., wasted) during that period of time

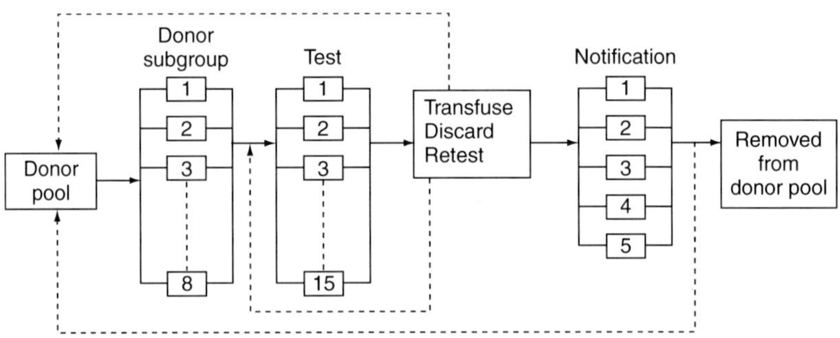

FIGURE 4-2 The decision support model for HIV antibody testing of blood and plasma donors. Source: Schwartz et al., 1990.

- the expected number of uninfected donors falsely notified
- the expected cost per donated unit for the particular screening regimen
- a wasted-unit index (the expected number of uninfected units discarded divided by the number of infected units discarded)
- the incremental cost and incremental number of wasted units for different screening regimens

This model was used at a meeting of an expert panel of the National Heart Lung Blood Institute to inform panelists who were deciding which blood-screening regimen to use at all blood-screening centers in the United States. The model provided the panelists with information and a basis for comparing screening regimens.

A collection of 15 screening tests was pared down to eight tests that could be used to detect HIV. The goal was to identify a sequence of tests that could reliably differentiate a unit of blood with HIV from a unit of blood without HIV. Given the results of a test, a decision was made to use, discard, or retest the unit. Decisions about notifying the donor were made next. If the donor was not notified, then he or she would be put back into the eligible donor pool. If the donor was notified of a possible HIV infection, he or she would be asked not to donate again, and his or her name might be placed in a donor registry of individuals who might have HIV infections.

The prevalence of the disease in high-prevalence areas, mainly in very large cities, was about 2.9 carriers in every 10,000 donors. In low-prevalence areas, the rate was about 1.6 carriers in every 10,000 donors. With an EIA test, initial sensitivity was about 0.98, with a range of 0.94 to 0.995. This range was entered into the model to find the scenarios that yielded the most reliable results.

During the first few weeks of HIV infection, it is very difficult to detect the virus, and it rarely shows up on tests. None of the tests modeled was good at detecting the very early stage of the disease. The most effective test had a sensitivity of 0.6, with a huge range of 0.11 to 0.83 and a negativity of 0.2, with another huge range. The specificity for this test was fairly good, but the range was 0.96 to 0.996. The model also provided estimated costs for performing different sequences of tests and for donor notification.

The model was used to examine various testing strategies with the objective of minimizing the risk per million donated units of accepting an HIV-contaminated unit of blood (Box 4-2).

BOX 4-2
Strategies for Screening Donated Blood for HIV

Strategy 1: All units are tested with an EIA: EIA-negative units are transfused; EIA-positive units are retested twice with the same EIA. If both subsequent EIAs are negative, the unit is transfused. If either of the subsequent EIAs is positive, the unit is discarded and a WB is performed. Donors with WB-positive results are informed of the results and placed on a registry.

Strategy 2: All units are tested with an EIA: EIA-negative units are transfused; EIA-positive units are tested with a WB. WB-negative units are retested with an EIA from a different cell line. All units positive by the first EIA are discarded, but only donors who test positive by WB or negative by WB but positive by the second EIA are informed of the test results.

Strategy 3: All units are tested with an EIA: All units are retested with the same EIA, regardless of the result of the first EIA. Units negative by both EIAs are transfused. If either EIA is positive, the same EIA is performed a third time. Units negative by two of the three EIAs are transfused. Units positive by two EIAs are discarded and tested with a WB. WB-positive donors are informed of the test results.

Strategy 4A: All units are tested with an EIA: EIA-negative units are transfused; EIA-positive units are retested with an EIA from a different cell line. If the second EIA is negative, the unit is retested again

Under the strategy used at the time of the study (in the 1980s), about 20.5 units per million in high-prevalence areas would be contaminated. In low-prevalence areas, about 4.7 units per million would be contaminated. In addition, almost 2,000 units of good blood from both areas would be thrown away because of false positives.

The best strategy for reducing risk to the population was Strategy 5; in high-prevalence areas 2.4 units of contaminated blood would be accepted in every million units, and in low-prevalence areas there would be 0.3 units per million. However, this strategy would mean that a large volume of good blood would be wasted.

with the second EIA. Units negative by two EIAs are transfused. Units with two positive EIAs are discarded and tested with a WB. Donors with WB-positive results are informed of the results and placed on a registry.

Strategy 4B: All units are tested with an EIA: EIA-negative units are transfused; EIA-positive units are retested with an EIA from a different cell line. If the second EIA is negative, the unit is retested again with the first EIA. Units negative by two EIAs are transfused. Units with two positive EIAs are discarded and tested with a WB. Donors with WB-positive results are informed of the results and placed on a registry.

Strategy 5: All units are tested with EIAs from two different cell lines. Units negative by both EIAs are transfused. All units testing positive by either EIA are discarded and tested with a WB. Donors with WB-positive results are informed of the results and placed on a registry.

Strategy 6: All units are tested with EIAs from two different cell lines. Units negative by both EIAs are transfused. Units positive on only one EIA are retested with the EIA that was positive and tested with a WB. WB-positive units are discarded and the donor is informed of the test results. Units positive by two EIAs but negative by a WB are discarded, but the donor is not informed of the results. Units positive by only one EIA and WB-negative are transfused.

Source: Schwartz et al., 1990.

Relevance to TBI Management

An mTBI version of this model could begin with field incidence, decisions then made, transfers, decisions then made, discharge, decisions then made, and so forth over time. Or it could begin with a later discovery of mTBI and the flows of decisions and therapies and probabilistic results.

To determine if a patient has an mTBI, he or she undergoes tests for, or may self-report, one or more of the following conditions or symptoms:

- a period of loss or altered level of consciousness
- a loss of memory of events immediately before or after the injury
- a change in mental state at the time of the injury (confusion, disorientation, slowed thinking, etc.)
- transient or permanent neurological deficit(s) (e.g., weakness, loss of balance, change in vision, praxis, paresis/plegia, sensory loss, aphasia, etc.)
- an intracranial lesion

Although tests may indicate mTBI, each test has initial and conditional sensitivities and specificities. Moreover, because there is no "gold standard" or best-practice diagnostic test, false positives and false negatives are common. Follow-on tests, such as magnetic resonance imaging (MRI), functional MRI, and other noninvasive procedures, also have initial and conditional sensitivities and specificities.

Because there is no best-practice therapy for the various levels and causes of mTBI, therapies also have a probabilistic chance of success or failure. In addition, the patient and family's involvement and compliance in therapies may modify the probability of a successful outcome. In the framework of the model, decisions about therapies or treatments are like the staged decisions described above about accepting or rejecting a unit of blood and whether or not to inform the donor.

The model is rich in possibilities for a decision-tree analysis of the timing and choices of tests, therapies, locations, personnel involved, and modes of care delivery (Figure 4-3). The model can also simulate different or multiple criteria to maximize the long-term quality of life for the patient and family and/or to minimize the long-term costs of treatment.

EXAMPLE 3
POLICY DECISION MODELING OF THE COSTS AND RESULTS OF MEDICAL SCHOOL EDUCATION

The purpose of this policy decision model was to support the decision of a board of regents and state legislature on financing state-funded medical education and meeting the state's long-term needs for physicians in rural and urban areas (Lee et al., 1987). Funding for medical

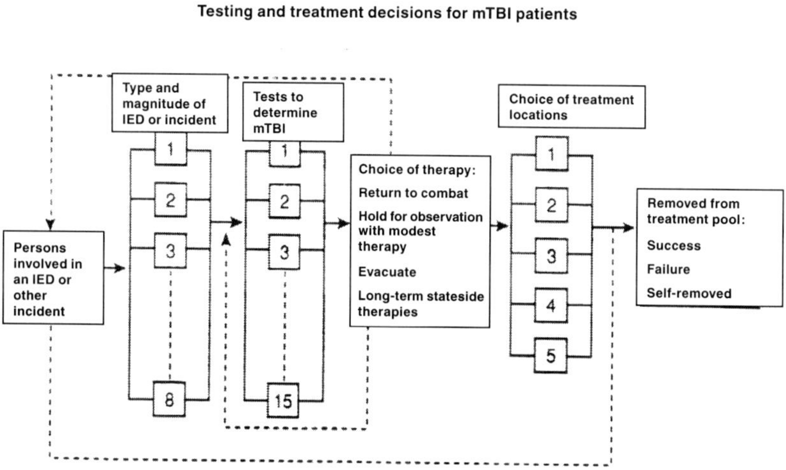

FIGURE 4-3 The decision support model for testing and treatment decisions for mTBI patients. Source: Adapted from Schwartz et al., 1990.

education in the state—which is not identified in the paper—was characterized as one of the highest in the country, although the state's per capita income was relatively low. In addition, most graduating physicians and medical residents left the state for more lucrative employment elsewhere.

The variables and parameters in the model were in-state, out-of-state, urban, and rural students and residents entering medical schools and hospitals in the state over time; all teaching resources necessary to educate medical students and residents (faculty, technology, facilities, and programs); financial costs of education; the practice locations and specialties of physicians who had completed their studies; the needs of citizens for physicians by location and specialty; and the goals and objectives of the regents and legislature for medical education in the state.

Possible policy scenarios were developed from input received from the state's board of regents (Box 4-3). The scenarios were evaluated with regard to how well each met the regents' goals and objectives.

Each evaluation was done using a decision-support model with four interrelated modules: (1) a teaching-resource model, (2) a financial-cost model, (3) a physician-output model from medical schools and residency programs, and (4) a physician practice location and specialties model (Figure 4-4).

BOX 4-3
System Goals Developed Using Nominal Group Technique with State Board of Regents

1. Develop an organized strategy for providing medical education in the state that is responsive to state needs.
2. Provide quality medical education within available resources.
3. Provide quality medical education in an efficient and effective manner.
4. Emphasize primary care training.
5. Provide a reasonable opportunity for state residents to obtain quality medical education.
6. Provide the appropriate number and type of physicians needed in the state, and encourage an appropriate demographic distribution of physicians.
7. Increase cooperation of M.D.'s and D.O.'s in education and services.
8. Improve the health of state residents.

Source: Lee et al. 1987. Reprinted with permission from INFORMS.

The teaching-resource model was a mathematical programming model. The financial-cost model was a detailed cost-accounting model. The physician-output model was a Markov-chain model. The physician practice location and specialties model was a stochastic forecasting model.

Based on the analyses, the board of regents and the legislature passed laws and implemented the following decisions:

- One of the three medical schools was privatized, and its state funding would be significantly reduced over time.
- Tuition was increased for all students (a larger increase for out-of-state students). In addition, a no-interest, revolving loan fund was established for students who practiced medicine in the state for a certain number of years, and a loan-forgiveness program was

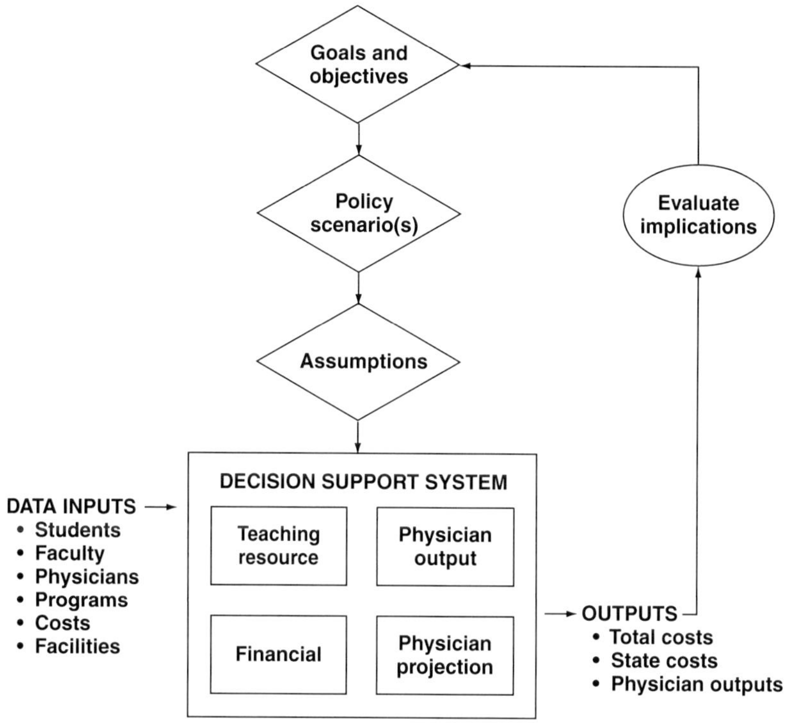

FIGURE 4-4 Strategy development and evaluation process. Source: Lee et al., 1987. Reprinted with permission from INFORMS.

established for students who remained in the state and practiced in rural areas for a certain number of years.

Relevance to TBI Care Management

The purpose of this model in the context of TBI care management is to illustrate that a modular OSE model can be used to evaluate various scenarios for achieving overall goals and objectives set by decision makers, even in complex health-delivery situations, such as those of the U.S. Department of Defense (DOD) and VA.

The modular, interconnected model might explore, for example, how TBI care should be organized on a regional basis. TBI care involves resources and resource planning, finances and costs, people (caregivers, patients, and families), and movement and availabilities, as well as forecasting of future needs and outputs. In short, TBI care involves many of the same elements as in the medical education model described above, including data needs of people, technologies, facilities, costs (fixed, semi-fixed, and variable), goals, objectives, and scenarios of interest.

The overall TBI model might be laid out in much the same way as the education model (Figure 4-5) but with names, flows, and data specific to TBI.

EXAMPLE 4
THE HEALTHCARE COMPLEX MODEL

The Healthcare Complex Model was developed for MHS in the 1990s and has been continually updated by Vector Research Inc. (now the Altarum Institute) in a prototyping process. The purpose of the model was to help MHS address a large number of capacity, organizational, resource allocation, and process change issues.

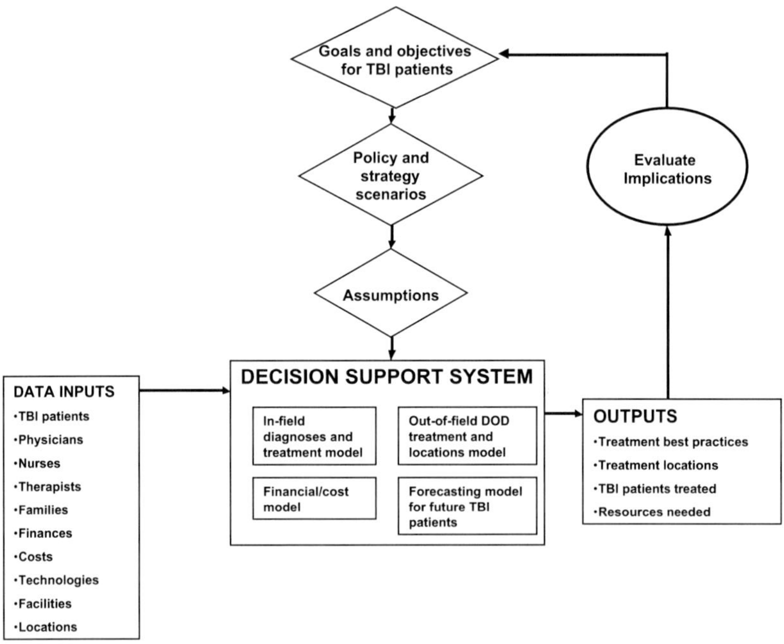

FIGURE 4-5 Decision-support model for developing a strategy and evaluation process for TBI care management. Source: Adapted from Lee et al., 1987.

As shown in Figure 4-6, the simulation represents a patient episode, from initial diagnosis through further diagnoses, treatments, monitoring, rehabilitation, and so on. The model shows the network of facilities at one or two major medical centers, a group of 5 to 10 hospitals, and multiple clinics. It also shows the dynamic input of patients who enter the system through clinics, hospitals, or major medical centers. Patients can be referred anywhere in the system and receive care by visiting a facility or care provider or by telemedicine.

The model simulates the dynamic flow of patients through a health care system as they are being diagnosed, treated, and rehabilitated. Their movement through the system, and even outside the system, is monitored throughout the episode.

The simulation includes about 60 providers and is driven by care protocols, in this case more than 1,200 ICD-9 code conditions (e.g., asthma, open-chest wound, low back pain) and roughly 1,400 to 1,500 peacetime disease-condition protocols. About 400 to 500 different protocols for deployment medicine, and several for dental care, move patients through the system as they receive treatment via the protocols and exit the system. The protocols involve care providers, ancillary personnel,

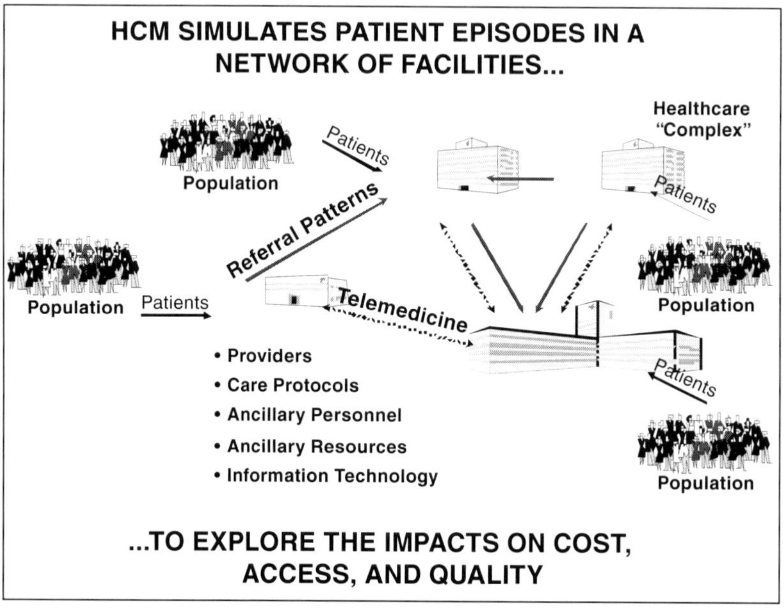

FIGURE 4-6 Overview of the Healthcare Complex Model. Source: Bonder, 2005.

and other resources and services, such as laboratories, x-rays, ambulatory surgery, pathology, pharmacy, and information technology.

This kind of model can address many strategic and tactical questions. For example, strategic questions might range from how many hospitals and/or clinics are required to the details of the kinds of protocols that should be tested. With a model of this kind, one can experiment with changes to a particular health care delivery system, as was done when it was applied to the Madigan Army Medical Center health care enterprise, which involved a simulation of a full year of health care delivery services at the facility (Chin et al., 2007). When the results were tested against the actual data, the dynamics were very, very close to those of the real-world system. This simulation demonstrated that OSE techniques can, in fact, simulate a large-scale enterprise of health care delivery to address a whole spectrum of issues.

EXAMPLE 5
A MIXED-INTEGER PROGRAMMING MODEL TO LOCATE TRAUMATIC BRAIN INJURY TREATMENT UNITS IN THE VA

Although the model described in this section was only mentioned briefly at the workshop, it represents a direct application of OSE to the management of TBI patients in the VA medical system. It is summarized here for illustrative purposes.

Côté and colleagues (2007) developed a mixed-integer programming "optimization" model to help the VA decide where to locate TBI treatment units in the existing system of VA medical centers. The goals were to minimize total patient costs for treatment, lodging, and travel, as well as costs associated with missing targets for the use of the services provided by the facilities.

The many variables and parameters in the model include the location of medical centers, patient locations and travel distances to medical centers, patient attendance attrition due to the distance and cost of travel and lodging, TBI severity levels, lengths of stay by severity, the size of medical centers, as well as their capacities and capabilities for TBI treatment, and the costs associated with these variables.

The model was designed to help with treatment location decisions in the six Florida-based VA medical centers in Veterans Integrated Services

Network Region 8 (VISN 8). Many scenarios were evaluated, such as opening one, some, or all six of the medical centers for TBI treatment at various levels. Each scenario was evaluated on the basis of admission retention rates, severity levels, and numbers of TBI patients treated at each center, expected treatment costs, expected lost-admission penalty costs, expected capacity penalty costs, and expected lodging and travel costs.

The results show that this model was useful for locating TBI treatment units, as well as for planning the development or consolidation of rehabilitation programs. The results and system outputs were extremely sensitive to the management structure and environment of the TBI treatment units in VISN 8 and suggested that careful consideration be given to the centralization of health care facilities and admission retention rates, as well as to interactions among these factors, when making decisions concerning the location of treatment facilities.

VA management could use the results to make informed decisions about the number, location, capacities, personnel, and costs of opening facilities throughout the region to treat TBI patients, with or without family members in attendance.

CONCLUSION

The five examples of OSE modeling discussed in this chapter illustrate the types of problems and objectives that quantitative OSE tools and methods have been used to address in a variety of health care-related settings. The examples showcase a number of technical approaches and how they are structured and illustrate data needs, critical assumptions, constraints, and metrics for evaluating performance using an OSE analysis.

The following chapter provides a detailed case study of the development, implementation, and sustainability of "system-supported" clinical practice at a major academic medical center—an approach that involves the integration of OSE tools and methods with information technology and a team-based approach to care delivery.

REFERENCES

Bonder, S. 2005. Changing Health Care Delivery Enterprises. Pp. 149–152 in Building a Better Delivery System: A New Engineering/Health Care Partnership, edited by

P.P. Reid, W.D. Compton, J.H. Grossman, and G. Fanjiang. Washington, D.C.: The National Academies Press.

Chin, S., R. Bannick, and A. Kress. 2007. Military Health System (MHS) Beneficiary Access to Trauma Centers. Available online at *http://gis.esri.com/LIBRARY/USERCONF/HEALTH07/DOCS/MILITARY_HEALTH.PDF.*

Côté, M.J., S.S. Siddhartha, W.B. Vogel, and D.C. Cowper. 2007. A mixed integer programming model to locate traumatic brain injury treatment units in the Department of Veterans Affairs: a case study. Health Care Management Science 10(3): 253–267.

Hazen, G.B. 2004. Dynamic Influence Diagrams: Applications to Medical Decision Making. Chapter 24 in Operations Research and Health Care, edited by M. Brandeau, F. Sainfort, and W. Pierskalla. Boston, Mass.: Kluwer Academic Publishers.

Lee, H., W.P. Pierskalla, W.L. Kissick, J.H. Levy, H.A. Glick, and B.S. Bloom. 1987. Policy decision modeling of the costs and outputs of education in medical schools. Operations Research 35(5): 667–683.

NAE/IOM (National Academy of Engineering/Institute of Medicine). 2005. Building a Better Delivery System: A New Engineering/Health Care Partnership, edited by P.P. Reid, W.D. Compton, J.H. Grossman, and G. Fanjiang. Washington, D.C.: The National Academies Press.

Schwartz, J.S., B.P. Kinosian, W.P. Pierskalla, and H. Lee. 1990. Strategies for screening blood for human immunodeficiency virus antibody. Use of a decision support system. Journal of the American Medical Association 264(13): 1704–1710.

5

Case Study Vanderbilt's Journey Toward System-Supported Practice[1]

William W. Stead

This discussion focuses on how to shift from expert-based practice to system-supported practice. To begin, one important observation is that if a unit or institution performs each of seven practices 90 percent of the time, the probability that it will perform all seven for an individual patient is only 48 percent. Health care providers and practitioners often do not realize this, however, because they report successes when they achieve 90 percent on individual practices. Overall, they may not achieve the desired clinical outcomes.

A shift from expert-based practice to expert-managed, system-supported practice can increase the level of success. A case example is the work done at Vanderbilt University for the past two years to improve ventilator management. This example, described below, clarifies how a clinical unit can deal with two challenges. Challenge I involves translating evidence into standard practice (Stead and Starmer, 2008). Challenge II is creating something that works, like closed-loop control, while keeping people in the loop (Stead et al., 2008).

In expert-based practice, experts are supposed to bring knowledge and technical skills, assimilate data, make wise decisions, and do what

[1]This chapter is based on the author's presentation and responses to questions during a plenary session of the NAE-IOM workshop on Harnessing Operational Systems Engineering to Improve Traumatic Brain Injury Care in the Military Health System on June 11, 2008. The author would like to thank NAE staff member Jessica Buono for her assistance in preparing this material for publication.

is necessary to ensure that their decisions are carried out. This practice is built around the expertise of the physician, and the performance of the system depends on the performance of this individual. Disagreement is expected among experts, and the performance of the system is no better than the performance of the individual expert. Typically it is worse, and experts are responsible for recognizing and learning from their mistakes.

In comparison, the idea behind system-supported practice focuses on the system's performance; teams of people, a well defined process, and information technology tools work in concert to produce desired results consistently. People bring compassion and judgment, the process brings simplification and standardization, and information technology reduces dependence on memory and forces action when needed. Collectively, the goal is to ensure the desired performance every time, and, if this fails, each failure becomes an immediate and iterative improvement. Figure 5-1 depicts this systems approach to health care, which joins system development to cycles of system-supported practice. The left-hand side represents iterative cycles of system development. First, a high-priority population is selected and defined, such as ventilator patients. The term population is used to imply a condition that needs to be managed to

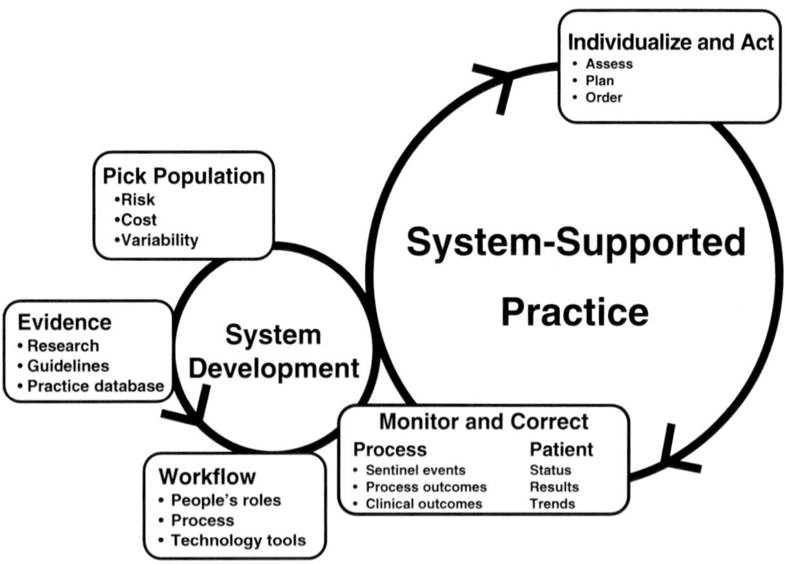

FIGURE 5-1 Systems approach to care.

achieve a consistent outcome. Although only one population is worked with at a time, a patient can be in multiple populations.

After selecting a population, an evidence base is gathered related to the target population, including literature on clinical research and consensus practice guidelines. Each consensus guideline is decomposed into a column of a table with a row for each practice it recommends. In the case of ventilator management at Vanderbilt, the table had a column for the University Hospital Consortium (UHC), one for the Institute of Healthcare Improvement (IHI), and one for the Centers for Disease Control and Prevention. The completed table displayed the variability in available evidence. A small group of subject-matter experts (two to six individuals) was then assembled to build a straw-person set of standard practices and an idealized process for executing them (Figure 5-2), as well as methods of measuring performance of the practice. A common fact base was compiled by documenting actual performance against the straw-person set of practices. This is the starting point for a model. At this point, the focus changes from a common fact base to cross-enterprise agreement.

In this instance, the focus was on intensive care units. Executives, medical, and nursing leadership from each unit, subject-matter experts, and key support personnel came together for an intensive day of cross-enterprise design. They used the common fact base developed from the straw person to identify problematic practices, processes, and measures. Then they broke into groups to work out designs for alternative solutions, reported their results, and agreed on revisions to the straw person. The design was limited to changes the group believed could be implemented in 45 days or less. It was also agreed that there would be a standard set of practices used throughout the entire medical center. For ventilation management, this meant these practices would be used from the neonatal intensive care unit through the burn and trauma units. These constraints eliminated many "blue sky" suggestions.

Through iterative cycles, evidence was translated into actionable standard practices. Take, for example, stress-ulcer prophylaxis. The UHC recommendation stated that mechanically ventilated patients are at high risk of developing gastrointestinal (GI) bleeding and should receive stress-ulcer prophylaxis, unless medically contraindicated. The IHI recommendation stated that stress ulcerations are the most common cause of GI bleeding in ICU patients and that the presence of GI bleeding is associated with a five-fold increase in mortality compared

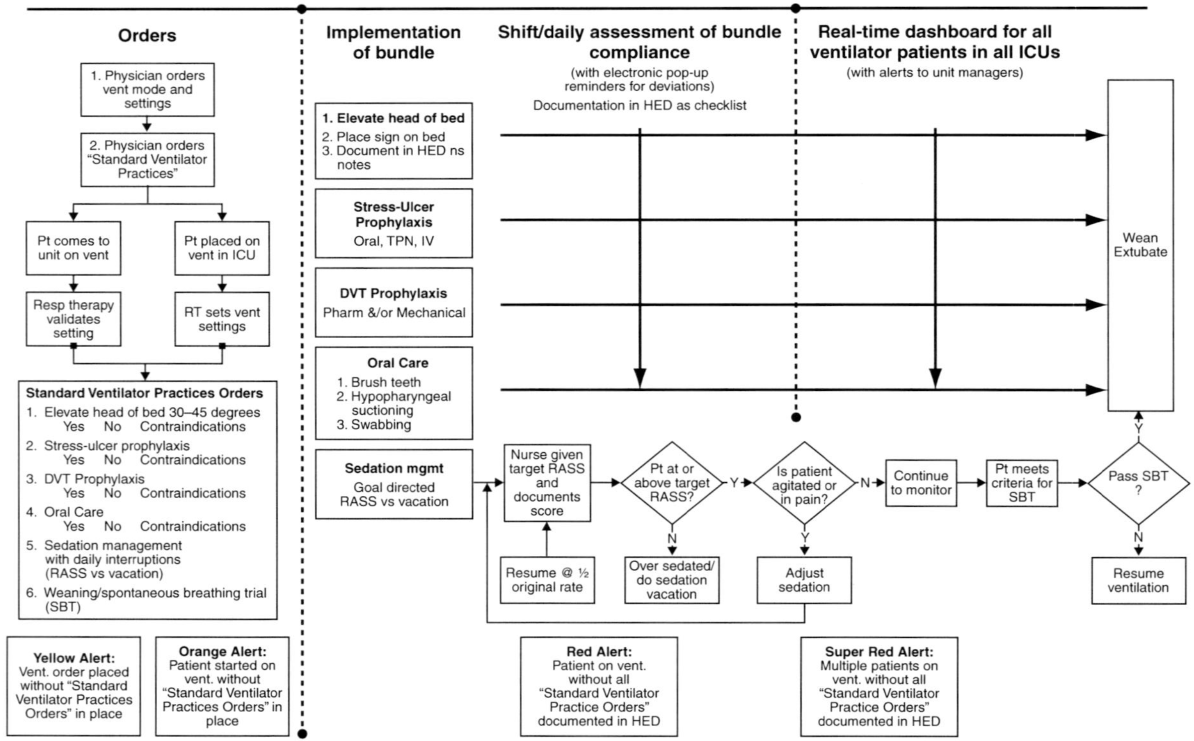

FIGURE 5-2 Supporting practice (process).

with ICU patients without bleeding. Therefore, prophylaxis is recommended. However, neither UHC nor IHI explicitly defined stress-ulcer prophylaxis, contraindications, etc. Table 5-1 shows how Vanderbilt eliminated the guesswork by outlining exactly which practice standard should be implemented in specific situations.

If parts of a recommended practice would be performed by different people at different times, they were subdivided into a standard practice for each role or time. For example, a recommendation for oral care was divided into three elements, including oral swabs, tooth brushing, and hypopharyngeal suctioning. Together, the elements of each standardized practice make this approach actionable and permit more focused accountability and performance measurement.

A major breakthrough in the development of system-supported practice at Vanderbilt was the process-control dashboard (Figure 5-3), which shows the entire team the status of a patient against their plan as a set of red, yellow, or green lights. There is a line for each patient and a column for each element of the standardized practice. Each cell represents one action that must be taken by a member of the team. No actions are lumped together. A green light means everything is in the expected state according to that plan. A yellow light means action must be taken, but there is still some time to do so. A red light means an outcome is out of line with the desired performance and action must be taken immediately. By making the display available in real time, the clinical team can tell what is right and wrong about the practices or documentation and help the system improve. Defining the system and outlining differences between evidence and standard practices are key aspects in the development of system-supported practice.

The second aspect of system-supported practice is closed-loop control, which means the output of the system feeds back directly to change

TABLE 5-1 Standard Practice Table for Stress-Ulcer Prophylaxis

VUMC Practice Standard:

Gastric Access	Gross Blood	
Y	N	H_2 blocker per tube
Y	Y	Proton pump inhibitor per tube
N	N	H_2 blocker IV or TPN
N	Y	Proton pump inhibitor IV

Bed	Patient name	Age	LOS	Orders				SBT				RASS					
							Vent	Scrn	Trial	DVT	SUP	Ord.	Pt.	HoB	swab	teeth	hySx
3002B	T, V W	72y	6 d		flowsheet	MAR	v	F		v	v	-4	-4	30			
3003X	N, D	60y	17 d		flowsheet	MAR	v	F		v	v	0	-2	45			
3004B	T, P L	64y	34 d		flowsheet	MAR	v			v	v	-1	-1	30			
3005A	C, D E	61y	7 d		flowsheet	MAR				v	v	0	-1	30	v	v	
3005B	B, J	66y	7 d		flowsheet	MAR	v	F		v	v	-1	-3	30			
3006X	W, A A	20y	66 d		flowsheet	MAR	v			v	v	-1	-2	30			
3007X	W, L E	49y	9:14		flowsheet	MAR				v		0	-1	30			
3008X	P, J L	69y	50 d		flowsheet	MAR	v	F		v	v	0	0	30			
3009X	R, C	72y	15 d		flowsheet	MAR	v	F		v	v	-1	-2	30			
3011A	P, J E	88y	9 d		flowsheet	MAR				v	v	0	0	45	v	v	
3011C	J, W D	69y	2 d		flowsheet	MAR				v	v	0	-1	30			
3011D	P, P J	55y	10 d		flowsheet	MAR	v	P	P	v	v	0	-3	30			
3011E	R, R E	74y	9 d		flowsheet	MAR				v	v	0	0		v	v	
3011F	N, E Y	55y	3 d		flowsheet	MAR				v	v	-1	0	30	v	v	
3012A	S, J D	56y	14 d		flowsheet	MAR	v	F		v	v	0	0	30			
3012B	R, M	63y	10 d		flowsheet	MAR	v	F		v	v	-2	-2	30			
3013A	N, B D	60y	8 d		flowsheet	MAR	v	F		v	v	-3	-2	30			
3013B	H, S M	66y	16 d		flowsheet	MAR				v	v	0	-1	30	v	v	

FIGURE 5-3 Process-control dashboard showing real-time feedback for supporting practices. Source: Stead and Starmer, 2008. Reproduced with permission from the Vanderbilt University Medical Center. Copyright 2007 by the Vanderbilt University Medical Center.

the inputs, thereby changing system performance and bringing it back under control (see Figure 5-4).

A simple example of this is the interaction between a thermostat and a furnace to control room temperature. There are many inputs to the room temperature; the furnace is just one of them. The thermostat is set to a desired temperature with a targeted control limit for how far above or below the desired temperature it can go. When the thermostat senses the temperature is falling below the control limit, it calls for heat, the furnace turns on, the temperature rises, and the thermostat approaches the upper-control limit and turns off the heat. Because it keeps a record of its immediate past performance, if it overshoots the next time, the thermostat turns off a little earlier until the temperature is within the control limits. If someone opens a window and changes the inputs to the system, the thermostat adapts to the change without reprogramming. The desired performance is achieved without programming complex interactions

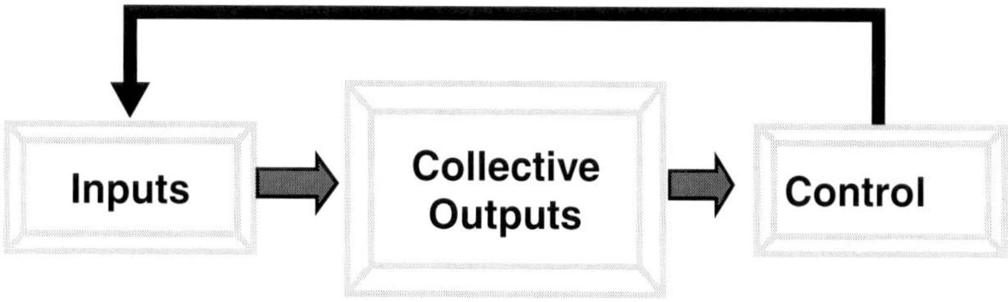

FIGURE 5-4 Closed-loop control. Source: Stead et al., 2008. Reprinted with permission of the American Clinical and Climatological Association.

among inputs or modifying the program as inputs change. This is what is needed in health care.

To achieve this, there must be agreement on an end-to-end plan of actions. Real-time measurement is also necessary to show what is happening and to display the instant status of the patient against the plan. The human team can then become the effecter mechanism that acts on input and allows for human judgment to override the system when the need arises. Figure 5-5 shows a model for adapting the practice for closed-loop control.

However, health care situations are often too complex for an end-to-end plan at the level of detail for an order or a prescription. Therefore, Vanderbilt developed a set of nested plans at different levels of specificity. Tier 1 is the highest level and has the least specificity. It outlines broadly

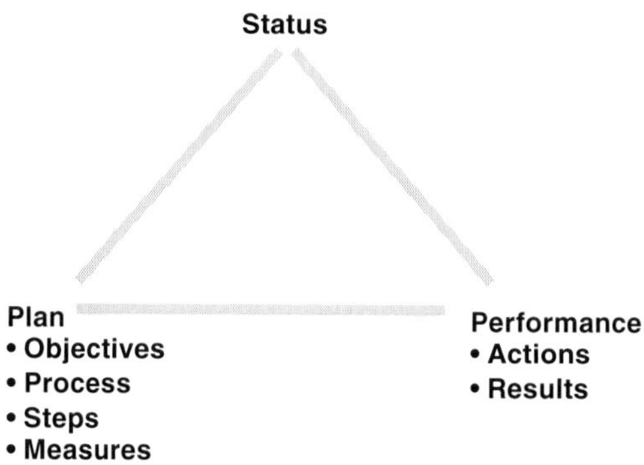

FIGURE 5-5 A model for adapting practice for closed-loop control. Source: Stead et al., 2008. Reprinted with permission of the American Clinical and Climatological Association.

applicable objectives and describes what is to be accomplished, rather than how to accomplish it. In the case of ventilators, the objective was to minimize the time a patient was on mechanical ventilation and, at the same time, minimize complications. Table 5-2 summarizes the highest level plan for ventilator management.

Tier 2 adds more specific plans and outlines how work is to be done to accomplish the Tier 1 objectives, including who should do what, how they should do it, and how it will be measured. Tier 2 is the level built into the cross-process control dashboards and tuned to the patient population and provider capabilities. Tier 2 must fit into the context of the Tier 1 objectives agreed to by stakeholders. Table 5-3 shows the Tier 2 plan for the objective of minimizing time on mechanical ventilation.

Tier 3 includes enough detail for caregivers to flex the plans in the tier above to the individual patient and to day-to-day variations in the facility's medication formulary, etc. Tier 3 is the level represented in order sets. Table 5-4 shows the Tier 3 plan for stress-ulcer prophylaxis. Overall, with these three tiers, the goal is to develop a modularized end-to-end plan that makes the complexity of any component manageable.

The time line for the Vanderbilt initiative is shown in Figure 5-6. All of the preparatory work was done in January 2007, and the design was completed in February 2007. By March, the order sets and education had been implemented. The process-control dashboard was developed and available to the team by May 2007, so, as they made rounds, they could see what was actually going on and what the dashboard showed as the status of their patient in comparison to their plan. They could

TABLE 5-2 Ventilator Management: Highest Level Plans (Tier 1)

Objectives	Process Steps	Measures
Minimize time on mechanical ventilation	• Avoid over-sedation • Wean as rapidly as possible	• Risk-adjusted time on mechanical ventilation • Unplanned extubation rate • Failed extubation rate
Minimize complications	• Pneumonia prophylaxis • Stress-ulcer prophylaxis • DVT prophylaxis	• Incidence per 1,000 ventilator days

SOURCE: Stead et al., 2008. Reprinted with permission of the American Clinical and Climatological Association.

TABLE 5-3 Ventilator Management Tier 2 Plans

Objectives	Process Steps	Measures
Goal-directed sedation	• Physician order for individual target RASS q12hr • Nurse assessment of actual RASS q4hr	• Ordered value, date, and time • Charted value, date, and time
Rapid weaning	• Rapid weaning protocol • If >24 hr daily screen for trial readiness by RT • If pass, spontaneous breathing trial within 6 hrs	• Screen pass/fail, date, and time • Trial pass/fail, date, and time

SOURCE: Stead et al., 2008. Reprinted with permission of the American Clinical and Climatological Association.

TABLE 5-4 Ventilator Management Tier 3 Plans

Objectives	Process Steps	Measures
Stress-ulcer prophylaxis	• + tube – blood => Pepcid 20 mg PT q12 hr • + tube + blood => Prevacid 30 mg PT q24 hr • – tube – blood => Pepcid 40 mg /c TPN q24 hr • – tube + blood => Pepcid 20 mg IV q12 hr	• Use or non-use of order set • Tube +/- • TPN +/- • Drug order • Administration record

SOURCE: Stead et al., 2008. Reprinted with permission of the American Clinical and Climatological Association.

easily report whether a practice shown as red was applicable, whether documentation was wrong, or whether the algorithm used to calculate the color for the dashboard was misleading. As a result, the process began to improve itself.

By September, the team felt that all of the processes were basically accurate, and they made it a high priority in every round of patient

February 2007	Bundle design shop
March 2007	Order set revisions & nursing education implemented
May 2007	Dashboard as Screensavers
June/July 2007	Dashboard refinements (resptherapy, oral care)
August 2007	Reports available
October 2007	"Improvement opportunity" reports available
Nov -December 2007	Emphasis on consistent execution of complete bundle for each patient
Jan -March 2008	Nurse charting improvements
March 2008	Renewed nursing education + educational materials

FIGURE 5-6 Time line for the Vanderbilt initiative.

contact to turn red cells into green cells on the process-control dashboard for each patient. Data received at the end of April showed a 50-percent reduction in the incidence of ventilator-acquired pneumonia in Vanderbilt Adult Hospital compared to the previous six-month period. This means 11 more people left the hospital, and 1.6 million dollars was saved in health care costs. In addition, with process improvement, Vanderbilt was performing all of the practices for 70 percent of patients and individual practices for more than 90 percent of patients.

A few key practices are highlighted in this model. The first is avoiding the desire to get everything right for everyone all the time. This is the wrong approach. The objective is to get something that is workable, deliver it, then constantly modify it so it will be self-correcting and self-sustaining. Table 5-5 outlines the rapid iterative cycles in the system-supported approach necessary to achieve a workable model.

The second key practice is to have more than one metric for success. The approach at Vanderbilt is measurement-driven based on agreement

TABLE 5-5 Rapid Iterative Cycles in the System-Supported Approach

Phase	Goal	# People	Duration
Initiation	Focus	2–4	1–2 hrs
Pre-work	Approach	6–10	2–6 wks
Design shop	Agreement	30–60	1–2 days
Development	Components	10–20	3–12 wks
Pilot	System	30–100	2–6 wks
Rollout	Dissemination	> 100	4–6 wks
Improvements	Performance	> 100	Continuous

on a standard set of practices and an iterative improvement process targeted at execution with 100 percent reproducibility. If continued measurement of a practice shows the desired improvement, it continues. If not, it is changed. Rather than relying on one metric, a set of indicators is used; each indicator has an explicit definition consisting of the process or system, the indicator, the actor, and the timing of the process.

Overall, system-supported practice is a combination of people, processes, and technology. It cannot be achieved by inserting information technology into our current way of working. The important thing is to focus on what must be improved rather than on external measures; use measurement-driven, iterative cycles to create self-correction and sustained improvements; use a common fact base to encourage agreement; set a target of 100 percent performance for the set of practices appropriate to a patient; and combine people, processes, and technology to achieve desired results. Another factor critical to the success of this initiative is that the institution and the support staff must approach the project with a collective will for process improvement.

REFERENCES

Stead, W.W., and J.M. Starmer. 2008. Beyond Expert-Based Practice. Pp. 94–105 in Evidence-Based Medicine and the Changing Nature of Health Care. 2007 IOM Annual Meeting Summary. Washington, D.C.: National Academies Press.

Stead, W.W., N.R. Patel, and J.M. Starmer. 2008. Closing the loop in practice to assure the desired performance. Transactions of the American Clinical and Climatological Association 119: 185–195.

6

Suggestions for Analysis Plans by Working Groups

During the planning phase of the workshop, the steering committee compiled a list of 36 issues raised by stakeholders in the Military Health System (MHS) medical community related to the care of patients with mild, moderate, and severe traumatic brain injury (TBI) (see Appendix B). From this list, they identified the areas that could potentially benefit from operational systems engineering (OSE) approaches and categorized them into five major challenges for TBI care management:

1. Development of new TBI knowledge
2. Detection and screening of TBI conditions
3. TBI care coordination and communication
4. Measurement and forecasting of demand for TBI care
5. TBI care system capacity, organization, and resource allocation

The committee then converted the stakeholder issues in these five categories into two or three issues for OSE analysis (i.e., analytical challenges that, if addressed effectively through OSE approaches, would answer important questions and help improve the performance of TBI care) (see Appendix C).

Forty of the 50 invited participants at the workshop (Appendixes F and G) were assigned, on the basis of their expertise, to one of the five working groups listed above.[1] The OSE analysis issues and challenges assigned to each working group are listed below:

[1]Steering committee co-chairs Norman Augustine and Denis Cortese and steering committee member Seth Bonder circulated among the five working groups.

- **Group A. Development of New TBI Knowledge**
  - A.1. Develop an approach to modeling the neuropathology and clinical dynamics of blast and concussive effects on brain function that lead to mild, moderate, or severe TBI.
  - A.2. Develop an acute-to-chronic disease model of mild traumatic brain injury (mTBI).

- **Group B. Detection and Screening of TBI Conditions**
  - B.1. Develop a model for diagnosing mTBI based on clinical experience and cognitive testing.
  - B.2. Develop the structure and processes of an mTBI screening program for use in theater and in the continental United States (CONUS).

- **Group C. Coordination and Communication for TBI Care**
  - C.1. Develop the structure of a TBI information system to track, monitor, and cue patients, families, and relevant providers.
  - C.2. Develop a methodology for coordinating the delivery of TBI care services immediately following trauma.

- **Group D. Measuring and Forecasting the Demand for TBI Care**
  - D.1. Based on historical data, develop a statistical estimate of TBI in the population of military personnel involved in Operation Iraqi Freedom/Operation Enduring Freedom (OIF/OEF).
  - D.2. Develop a methodology of forecasting the time stream of future TBI cases in the military population.
  - D.3. Develop elements of an assessment and a methodology for assessing the value of preventing TBI.

- **Group E. Capacity, Organization, and Resource Allocations for a TBI Care System**
  - E.1. Describe elements, processes, and activities to represent the dynamics of a complete course of TBI care as input for a model of a TBI care system.
  - E.2. Outline the structure of a model or methodology to assist in planning for the allocation of scarce TBI care providers in theater and in CONUS.

For each challenge, the working groups were asked to design a suggestion for an analysis plan, including an objective, a technical approach, an approach structure, data requirements, critical assumptions and constraints, metrics, expected output, implementation actions, estimates of time and resource requirements, and other elements of a future OSE study or program that might be initiated to meet the target challenge. Specifically, working groups were asked to identify the types of approaches, methods, and information that could be developed by OSE practitioners to assist care providers and managers in delivering quality TBI care.

Each group had a chairperson who served as the technical lead, a rapporteur who was responsible for summarizing and communicating the technical aspects of the group's approach, and several experts in relevant subject matter. The group was asked to review its assigned tasks; modify them as appropriate; and develop analysis plans for a study, method, or means of data collection. Each group was also instructed to design its analysis plans so they could potentially be used by MHS. Group members offered individual suggestions and ideas for the development of the analysis plan, and no attempt was made to ensure a consensus. The chairman and rapporteur of each working group presented a summary of the group discussion to the full workshop. The five working groups were not Academy-appointed committees, and this summary reflects the views of the individuals who participated in each working group—not necessarily those of the institution or the workshop planning committee.

The analysis approach to each task comprised three parts: the issue itself (what each group was asked to do and the purpose of the task); the reasons the task was addressed (in terms of the stakeholder issues on which it was based); and the analysis output (the study, approach, or method that could be developed by implementing the analysis plan, essentially describing the capabilities of a specific model based on the suggested approach).

Each plan for a future OSE study or program included an objective stating what could potentially be achieved through this approach; a description of the technical approach to achieving the specified objective; and an approach structure, including the variables considered for inclusion, the development of necessary relationships, important statistical and structural model formulations, and a description of how the technical approach would be implemented.

In addition, each group was asked to identify data that might have to be collected, as well as critical assumptions about future demand and data availability and critical constraints based on available measurement tools, sample sizes, and safety regulations. Each group also outlined metrics for evaluating performance, outcome, and utility and identified specific tasks for executing the suggested approach. Finally, each analysis plan included expected output, implementation actions, estimates of duration, and resource requirements.

The following summaries of the suggestions for analysis plans are based on presentations by the chair and rapporteur of each group. It is important to reiterate that the analysis plans are suggestions developed by the working groups for the express purpose of illustrating potential applications of OSE tools and methods to a select sampling of specific TBI care challenges. The analysis plans should not be construed as consensus recommendations of the individual working groups, the workshop participants as a whole, or the National Academies.

WORKING GROUP A: DEVELOPMENT OF NEW TBI KNOWLEDGE[2]

The phenomenology surrounding TBI, particularly mTBI, is not completely understood. This limited understanding also limits objective diagnosis, the effective management of symptoms at the point of injury, and appropriate acute patient care. In addition, because little is known about the progression or disappearance of mTBI symptoms over long periods of time after exposure to a blast injury, the effectiveness of triage, rehabilitation, long-term disease management, and efficient use of community services for mTBI patients are all limited. Thus improving the understanding of TBI injuries and mechanisms is a crucial issue for TBI care providers and managers trying to develop effective treatment protocols.

The two specific tasks assigned to Group A both focus on the development of new TBI knowledge. The first involved developing an approach to modeling the neuropathology and clinical effects of blast and concussive injuries on brain functions leading to mild, moderate, and/or severe TBI. The second task was to develop an acute-to-chronic

[2]See Appendix G for names of working group members.

disease model of mTBI showing the evolution of disease states or symptoms over time for a population of mTBI patients, some who were initially asymptomatic and some who were overtly symptomatic. Because of the need for better data collection on TBI to inform future research, the group also addressed a third issue, the establishment of a national database on brain trauma events in the civilian population.

Issue 1: Methods of Measuring Brain Vital Signs

Issue 1 focuses on the development of methods of measuring brain vital signs in patients with blast and/or concussive injuries, that is, patients with TBI from blast, blast plus concussion, and concussion alone. The suggested technical approach includes animal and human studies using neuroimaging, neuropsychological testing, and neuropathology assessments to (1) measure temporal changes in the brain in vivo after a blast and/or concussive event and (2) identify biomarkers, brain edema, changes in brain blood flow and volume, and changes in neurochemistry. Once these brain changes and biomarkers have been identified, care providers can determine the efficacy of potential treatments, specific risk factors for the development of TBI, and effective prevention measures.

One of the main challenges in treating mTBI is identifying who is affected and how their outcomes develop over time. Therefore, a necessary data requirement for this initiative would be pre-deployment neuropsychological screening to obtain baseline data on a sample of potential patients. The baseline data could then be compared to other pre-deployment and post-deployment tests for changes in brain chemistry. Data would also be used to evaluate deployable, alternative, and inexpensive methods of measuring brain vital signs, including blood flow, blood volume, and brain edema.

A serious problem in post-screening at various points in theater is patients' denial of injury and the absence of symptoms. To address this problem, a research station could be established to analyze a group of patients and gather data prospectively at a Level II facility. Imaging research stations would also be beneficial at Level III, IV, and V MHS facilities and Department of Veterans Affairs (VA) facilities. These research stations would be capable of performing structural and functional magnetic resonance imaging (MRI) to measure brain anatomy.

To gather baseline data on asymptomatic individuals exposed to blast for comparison with controls, each vehicle could carry a "black box" equipped with sensors for characterizing the magnitude and other parameters of a blast. Individual soldiers could also be equipped with black boxes that could characterize the magnitude and specific effects of a blast on that particular soldier. The black boxes would collect necessary background information that is not otherwise available. Overall, gathering data on imaging and neuropsychological testing at early stages would make it possible to monitor and compare the effects of an injury event over time and provide a link between vital signs and estimates of event severity.

Data are also needed for analyses of systematic gross pathology relative to a patient's history, including information on previous TBIs, alcohol use, and mental health status. Information for such analyses might be collected from studies currently under way at academic and government research sites.

Critical assumptions associated with the analysis plan described above include the capability of MRI at one echelon (at least) in theater; standardization of imaging and support of data collection at Level II through V facilities; and the availability of a technically competent staff. In addition, it is assumed that data will be available on human neuropathology, potentially from evaluations of data related to mortality during the Global War on Terror and that animal studies will accurately mimic the human condition.

Critical constraints include obtaining military permission for in-theater research and access to postmortem neuropathological data and establishing a data repository to support collection and long-term analysis. MRI, position emission tomography, and single photon emission computed tomography would also be necessary through Level V at TBI centers in CONUS. Other critical constraints are the execution of research proposals, the relevance of animal models to the human condition, the feasibility of using large animal models to test neuropathology, and the variability of image interpretation among research staff.

Performance, outcome, and utility metrics would include (1) the validation of animal models via human imaging studies and (2) the tracking of patients to ensure adequate throughput of those imaged with mTBI at Level II facilities, as well as those imaged with moderate to severe TBI at Level III through V MHS facilities and VA facilities. It would be necessary for one Level II facility to gather information on a group of patients and a group of control patients, with a minimum of 30 patients per group. Although clinical analyses of MRIs often miss cases of mTBI, they still have some clinical utility because they provide data on ventricular volume and atrophy and can identify brain bleeding. This information is only available at Level II facilities if they are equipped with MRI imaging capabilities. In addition, neuropsychiatric testing could identify cognitive deficits and could thus be another metric for identifying health outcomes following TBI.

Tasks to execute the approach described above would include convening a working group of neuropathologists to address questions about TBI and associated exposures and outcomes. Issues to consider include differences between concussive and blast injuries, appropriate measures for these two types of injury, and determining if there is consensus on the sequence of neuropathology after a blast trauma. Tasks would also include establishing a database of results from neuropathological studies and ongoing experiments at government research laboratories and academic institutions. In addition, research proposals would have to be developed to support the proposed animal experiments.

Expected outputs include a determination of the differences between mild blast, mild concussive, and mild blast plus concussive injuries as revealed by MRI and neuropsychological measures; the classification of characteristics of mTBI patients who need additional medical surveillance and/or treatment; and the identification of temporal characteristics of blast/concussive exposure and longitudinal outcomes.

For a single Level II site to implement this analysis plan, Group A estimated that the project would last approximately 18 months and would cost about $3.5 million for in-theater equipment. Data collection would begin four months after initiation of the project and would require three active-duty personnel and two civilian personnel. Additional support resources (e.g., security) would also be required.

Issue 2: Anticipating Downstream Consequences of TBI

The objective of Issue 2 is to provide a means of understanding and forecasting the downstream consequences of TBI, including both the later effects of immediate post-trauma treatment and interactions with subsequent TBI events or other psychological traumas. The technical approach to meeting this objective involves modeling a finite state-space stochastic process in which the initial conditions are TBI incidents (timing and conditions), co-morbidity, and treatment. This model would also take into account instances with no co-morbidities and cases in which no treatment was administered because the injuries were not reported.

Necessary data would be collected on anomalies and triggering events and their associated consequences as a basis for developing a list of behaviors and types of triggering events that might be of concern in patients with TBI. Longitudinal data from a test sample of long-duration TBI victims (e.g., from the Vietnam Head Injury Study)[3] would also be useful, along with data on short-duration victims (e.g., from Iraq and Afghanistan). In addition, new surveys would be designed and administered to capture the effects of long-duration TBI. Overall, the purpose would be to identify medical problems and behaviors of concern, such as neurodegeneration and adjustment disorder, that are likely to be manifested in 1, 5, 10, 20, or 30 years, as well as triggering events that may complicate TBI and/or accelerate the emergence of those conditions.

Data collection would be focused on generating the structure of relevant TBI events to define the states of the system. (The state space would reflect the progression of the TBI patient from acute through chronic stages, and different states would reflect the treatment provided.) The data would redress our current lack of data, and hence our inability to define states and probability distributions of their occupancy times, as well as transitions between states. The initial data collection would involve analyses of patient records and 50 to 100 lengthy interviews for each group, followed by a more formal analysis instrument for samples of 1,000 to 2,000 patients. All data collection would be supplemented with meta-analysis and data mining, with the expectation that the results could suggest appropriate clinical trials of alternative treatments.

The overall goal is to establish a baseline frame of reference for detecting changes in the state space. Thus the approach for this analysis plan would involve enumerating the characteristics of a TBI event, co-morbidities, treatment, and baseline characteristics of each victim. Similar documentation of downstream consequences and potential physical and psychological sequelae would also be collected as a basis for making correlations among characteristics of the initial event, downstream consequences, and subsequent triggering events.

---

[3]National Naval Medical Center. Ongoing. Vietnam Head Injury Study (VHIS) Phase III. Available online at *http://www.bethesda.med.navy.mil/professional/research/vietnam_head_injury.aspx* (accessed September 29, 2008).

These correlations would require long- and short-term longitudinal histories of TBI patients from Vietnam and Iraq and Afghanistan, as well as appropriate control populations. The Vietnam population could provide an opportunity to examine long-term outcomes not yet realized in the Iraq and Afghanistan population; thus a cohort that was clearly identified as having been exposed to blast injuries in Vietnam would be selected to provide life histories and determine relevant outcomes since their exposures. Data on the forensics of TBI events, including relative location of the blast to the soldier and the type of vehicle the soldier was in when the blast occurred, would also be necessary to define the state at the conclusion of the event.

Critical assumptions for this analysis plan include (1) data on TBI victims will be available as a basis for identifying appropriate initial interview samples as well as subsequent samples for longitudinal analyses, (2) sufficient patient recall to identify events of interest, and (3) adequate reporting of the salient events. No critical constraints for this analysis are anticipated other than the availability of funding.

Metrics include (1) quality of life and its relationship to the events of interest, (2) treatment burden and costs, and (3) downstream needs for patient monitoring and response. Tasks necessary to execute this approach include (1) developing a proposal for the research, (2) securing adequate funding, (3) obtaining an interview sample, (4) executing interviews and preliminary analyses to generate the state space, (5) demonstrating an initial model based on the sample, (6) developing a formal instrument for data collection, and (7) using this instrument to prepare a model for initial application to TBI care.

Expected outputs include (1) identification of potential warning signs of downstream consequences of TBI, (2) identification of immediate actions and treatment choices to minimize downstream consequences, (3) classification of a pattern of downstream consequences following a TBI event and its treatment and costs, and (4) comparison of long-term treatment strategies. It is estimated that this project would last four years and require approximately $1 million in funding, with initial operational results expected 18 months from the start-up date of the project.

Issue 3: Database on Civilian TBI Events

Issue 3 was to create, in cooperation with federal and state agencies and private organizations, a nationwide database of information on the effects of crashes, explosions, and other traumatic events on the health of the civilian population over time. Data on head injuries from brain trauma events in the civilian sector, many of which may require treatment similar to treatment of mTBI on the battlefield, would be made available for analytical and comparative purposes to provide insights into traumatic events on the battlefield. Explosions in chemical plants, natural-gas pipelines, grain silos, car crashes, and other events generate blasts that have many characteristics in common with munitions explosions. There are also important differences that may be significant in recognizing the effects of battlefield injuries.

The national database would be continuously updated and made available for analysis by all interested parties. Data on the nature of brain trauma events, the surrounding conditions, the effects on personnel, and medical diagnoses of immediate and long-term effects would be required for implementation of this approach.

A critical assumption is that data from brain trauma events in the civilian sector would provide information important to improving the understanding of TBI effects in the military population. A critical constraint is that there is no central federal or civilian agency that could collect and archive data of this nature, let alone data detailed enough for analysis.

Performance, outcome, and utility metrics include identification of the characteristics of brain trauma events, primary clinical effects on humans, primary effects on brain structures and their impacts on humans, and long-term clinical effects on humans. To execute this approach, a federal program would have to be identified that would be responsible for receiving and archiving data, and a mandate or incentive system would have to be implemented to ensure that all traumatic brain events in the civilian sector were promptly reported to this agency. The expected output would be a large civilian database that MHS could use to augment its analysis of TBI and make comparisons with data collected in theater.

Implementation of this approach would require identifying one or more advocates for the creation of the proposed database, as well as the development of an organizational structure to support the activity. Although the costs cannot be estimated at this time, this would be an ongoing effort, and resources to accomplish it would have to be carefully explored. The U.S. Department of Transportation, which collects data on vehicle crashes, and perhaps some other agencies or organizations, could provide some support.

WORKING GROUP B: DETECTION AND SCREENING OF TBI CONDITIONS[4]

MHS needs better testing (cognitive tests, brain scans, etc.) for TBI, particularly for mTBI. The test results and other information could be used to develop an effective, efficient screening process that takes into account Type I (sensitivity) and Type II (specificity) detection errors. In the case of mTBI, physical symptoms may not signal the extent of the injury. Moreover, the current screening system relies on self-reporting of TBI causative events or symptoms in theater or at the end of a deployment. This system has been less than effective because individuals have multiple disincentives for self-reporting both during and after a deployment.[5]

The costs of false positives and false negatives are significant. mTBI "false alarms" remove healthy soldiers from service at significant cost to the military mission. However, if undetected, mTBI can impair job performance during a deployment, putting the individual, other soldiers, and the mission at increased risk. If not detected and treated promptly, an initial mTBI increases a soldier's risk of subsequent TBI and/or may result in long-lasting symptoms post-deployment that will degrade his or her quality of life.

Unfortunately, little objective information is available about the onset and progression of mTBI that can be used to assist in detecting and screening for mTBI injuries. There is, however, a good deal of subjective information (e.g., medical experience; neurological, cognitive, and psychological testing; imaging; questionnaires), as well as data on the incidence of mTBI, that could conceivably be used as an interim diagnostic vehicle for assessment, detection, and screening programs and in making "return to duty" (RTD) decisions.

Working Group B focused on two issues: (1) the development of systems engineering models for evaluating and improving the current TBI screening process, particularly for mTBI; and (2) the development of a predictive diagnostic model for mTBI (i.e., a means of estimating

---

[4]See Appendix G for names of working group members.

[5]Disincentives are based on the stigma associated with the injury; soldiers' reluctance to risk being removed from their units; leaving their comrades in arms short-handed for an invisible or (mis)perceived "minor" injury; soldiers' desire at the conclusion of a deployment to "just get home" and not prolong the post-deployment evaluation; and soldiers' fears of the negative implications of a positive diagnosis for long-term military careers.

the probability that an individual soldier returning from the field [or remaining in the field] has mTBI). The group focused mostly on developing an analysis plan to advance the first objective, which appears to be the more difficult technical (i.e., modeling) challenge. Less consideration was given to the second objective, which would be easier to model.

Issue 1: Evaluate and Improve the Current TBI Screening Process

In the first analysis plan developed by the working group, the expected outputs would be an improved understanding of how well the current screening and diagnostic processes work and mathematical models that suggest opportunities and requirements for improving them. Specifically, the models developed in the course of the proposed study would (1) recommend a staged approach to the screening instruments that should be used, specifying when, in what sequence, and in what locations (e.g., in-field, on the base, post deployment) they should be used, based on estimated risks, sensitivities, specificities, and costs; (2) evaluate the costs and benefits of the new screening tool, technology, or approach; and (3) assess usability and compliance issues associated with the screening and diagnostic process and the benefits of improving upon these.

The working group identified five complementary technical approaches: (1) determine problems or shortcomings of the current approach; (2) identify other approaches; (3) develop descriptive models of the current process; (4) develop prescriptive models for optimizing the screening process; and (5) develop learning models through data-mining techniques. Whatever models are developed, they must be able to accommodate and/or investigate data uncertainties (e.g., unknown costs, sensitivities, compliance).

One approach to determining the limitations of current processes would be to use mapping based on interviews, observations, focus groups, and surveys. Once these data are obtained, usability testing could be conducted, as well as analyses of cognitive tasks and cognitive work and analyses of training and implementation requirements.

Another approach would be to identify alternatives to the current approach, either by designing and evaluating (rapid prototyping) better methods or tests for screening or by looking to other domains for ideas. For example, investigators could study how athletes, miners, and industry workers at risk of TBI are tested.

A third approach would be to quickly develop descriptive models to determine how the current processes perform. These models could include probability models formulated as a Markov process, Monte Carlo models, cognitive-task analyses, judgment models (to identify the key cues used by experts), resource-requirements models (to identify the screening load), and throughput models (see Appendix H).

In addition, prescriptive models could be developed to optimize the screening process. These could draw upon classic optimization models (such as those described in the following paragraph), Markov decision processes (MDPs), partially observable MDPs (which lend themselves to handling uncertainties or partially observed data), and simulation models. Based on prescriptive models, one could predict the performance (e.g., cost, number of true and false positives, number of true and false negatives) of a given screening process and then adjust details of the process to improve performance.

Another conventional approach to addressing uncertainties would be to conduct parametric analyses and sensitivity analyses on models, for example, to determine which of the data elements initially targeted for collection had the strongest effect on the model results and, therefore, require the most improvement, perhaps by further data collection, analytical studies, and so on. Finally, signal-detection theory models could be used for making a diagnosis in a situation in which one must distinguish between signal and noise and determine an optimal criterion for classifying an individual as positive or negative for mTBI.

The group identified several objectives for prescriptive optimization models. First, a model could be developed to minimize the total costs of the current screening process, including both the direct costs of screening and the costs associated with the potential harm from false positives and false negatives. Another objective could be to minimize the expected number of tests and screens administered. A third objective might be to minimize the number of misses while controlling for the number of false alarms.

Classically, one could minimize total costs by including known direct costs, the costs of false negatives, and the costs of false positives and then conducting a sensitivity analysis. However, many of these costs cannot be clearly identified or quantified. Another way to model problems such as these is to remove the costs from the equation entirely and aim for a certain performance level with minimal screening costs. Because all of these objectives are likely to be important, however, one could formulate models that assess multiple objectives or criteria and then try to optimize the process with respect to multiple competing criteria. A multi-criteria optimization model, for example, might enable one to assess trade-offs between minimizing total costs, minimizing harm, and minimizing false negatives.

Note that the same general modeling approaches could be used for deployment and post-deployment screening scenarios.

Issue 2: Develop Predictive Diagnostic Models for mTBI

The second issue discussed by the working group was the development of predictive diagnostic models, a means of estimating the probability that an individual soldier returning from the field will have mTBI. Similar to the multiple-model approach described above, the proposed technical approach is to predict the probability of mTBI using several modeling approaches and to compare these approaches to determine which ones are most effective. Suggested approaches include basic probability models, logistic regression models (which assume good data on an individual's deployment history and TBI status), Bayesian networks, influence diagrams, and fuzzy logic models (the latter three can work with imprecise information; indeed, fuzzy logic models have already been used for predictive modeling of TBI in civilian populations) (Guler et al., 2008) (see Appendix H). Each of these approaches is fairly straightforward. The model developer would choose the one most likely to be useful based on process particulars and currently available data, following the general development approach described below.

The modeling approaches listed above should be treated as complementary and iterative. Given our limited understanding of the sensitivity and specificity of the current screening process, one could begin by developing a simple probability model of the current process and use it to conduct "what if" analyses to get a better understanding of how well the system performs. Later, the model could be expanded into formal optimization frameworks, but typically a great deal can be learned about how to improve outcomes by using the model to conduct simple trial-and-error inquiries.

In other words, one would develop a model, validate it, start to use it to identify improvements, and then build an optimization structure around it. Because the first model is often not exactly on target (either because it does not capture important aspects of the problem or because it requires unobtainable data), successful modeling almost always follows an iterative process. In this context, multiple types of models are often simultaneously prototyped to get an idea of which ones will be most useful. Usually no single model is chosen, but a few models are used together in complementary ways to address or illuminate different aspects of the problem.

Following this strategy, one approach would be to fund exploratory work, through academic grants, consulting, and/or seed funding for a number of short-term pilot studies (approximately 10 projects of 6 to 12 months duration), possibly focused on a subpopulation. Funding recipients would spend the first 6 to 12 months developing an in-depth understanding of problem specifics and model requirements, undertake preliminary data collection or knowledge elicitation, and build and evaluate preliminary models.

If seed funding were provided, successful prototype tests could lead to further funding for longer term studies (one to three years) to refine and expand the most promising approaches. This might be followed by a one- or two-year rollout on a larger scale.

Data requirements for the proposed research would include potential input variables, output performance data for model validation, and data for evaluating the effectiveness of system redesigns. Screening models would require data on different kinds of screening tests and methods, their costs and risks, their false negative and positive rates, the costs of false negatives and false positives, transition probabilities, compliance rates, resource requirements, and the time required to run a test and reach a decision. Data would also be gathered on the overall screening process so that when different screening tests are integrated into a more complete screening process, the effectiveness of the whole process can be measured. Data for model validation include output data, such as unit effectiveness (e.g., readiness, morale, cohesiveness in action) and other militarily relevant information.

For diagnostic models, data needs may include predispositions (e.g., genetics, physical and mental factors), past incidences (e.g., number, frequency, severity, consequences), and exposure (e.g., role in the armed forces). Further data requirements will constitute a much longer list, some of which would be identified in the development of the pilot models described above. Other information needs might include policy requirements or requirements based on a more in-depth understanding of problems with the current screening process, such as, for example, the Post-Deployment Health Assessment and the Post-Deployment Health Reassessment.[6]

In general, several pre-test/post-test comparison measures would be useful inputs to either type of model. Many types of data might be collected about the population to help clinicians infer whether mTBI is present in an individual, such as who patients are, where they have been, current and past diagnoses and treatments, baseline mental states, and so on. Evaluation metrics might include RTD, unit effectiveness, suicide rates, total costs, and the effectiveness of screening for the detection of TBI and related problems and the ability to distinguish between them.

Critical assumptions for success are that the data necessary for building models are available and that policies can be changed if they interfere with optimal decision making. Critical constraints include funding, the availability of non-financial resources, and competing priorities. Other constraints include the need to train people, the deployment of the screening process, tracking requirements and compliance, and building and maintaining the data infrastructure to support the improved screening process.[7]

WORKING GROUP C: COORDINATION AND COMMUNICATION FOR TBI CARE[8]

The primary charge of Working Group C was to develop the structure of a TBI information system to track, monitor, and cue care delivery for all TBI patients, no matter what the severity of their injuries. The

---

[6]Enhanced Post-Deployment Health Assessment (PDHA) Process (DD Form 2796). DOD Deployment Health Clinical Center, Walter Reed Army Medical Center, Washington, DC. Available online at *http://www.pdhealth.mil/dcs/DD_form_2796.asp* (accessed September 29, 2008); Post-Deployment Health Reassessment (PDHRA) Program (DD Form 2900). DOD Deployment Health Clinical Center, Walter Reed Army Medical Center, Washington, DC. Available online at *http://www.pdhealth.mil/dcs/pdhra.asp* (accessed September 29, 2008).

[7]These requirements and constraints were drawn directly from the DVBIC Working Group on the Acute Management of Mild Traumatic Brain Injury in Military Operational Settings, Clinical Practice Guideline and Recommendations, dated December 22, 2006, which lists many key elements only some of which are called out here.

[8]See Appendix G for the names of members of all working groups.

system would be useful for clinical monitoring and follow-up and would be accessible to, and would cue, all patients, patient families, and other relevant providers in the MHS, VA, and civilian sector. In addition, the group was asked to develop a methodology for the coordinated delivery of services for TBI and related co-morbidities immediately following trauma exposure. The methodology was required to take into account the needs and preferences of the patient and family members, as well as the resources (e.g., number and type of providers available) and infrastructure of the relevant health care system.

The objective of the analysis plan is to address DOD's lack of a system-wide approach for tracking and monitoring TBI patients to ensure the effective management of their complete care. Problems in the coordination of care have arisen between the MHS and VA and among all care providers at different levels of care and at different medical facilities (Cope et al., 2005; Sayer, 2006). The analysis plan and methodology described by the working group would also improve the timeliness, coordination, and efficiency with which TBI care resources (e.g., care providers, equipment, materiel, supporting organizations, and infrastructure) are brought together to address the needs of TBI patients during the first two or three days after the initial injury. Improving the operational responsiveness and coordination of the TBI care system will also improve patient outcomes and the efficiency of using scarce resources.

The desired output of implementing the analysis plan would be a proactive information system that facilitates the tracking, monitoring, cueing, coordination, communication, and scheduling of care for TBI patients from "cradle to cure." Such a system would also ensure that information flows and flows of care are aligned to provide effective, timely status awareness and response capability for TBI patients. An additional output would be an operational model and/or process methodology that could be used for the real-time allocation of TBI resources, thus ensuring that the delivery of clinical services is coordinated and responsive to patients' needs.

In working its way through the development of an analysis plan, the group's initial discussion focused on key considerations, such as that many military personnel who experience mTBI are not identified and, therefore, are not tracked or treated. Sometimes the manifestation of TBI symptoms is delayed. Sometimes, even if detected, TBI symptoms are confused with other medical conditions or are ignored by the soldier or his or her unit because of the urgency and gravity of the military mission. Moderate and severe TBI are easier to identify and to differentiate from other medical conditions. However, even severe TBI may be initially overlooked in the presence of other traumatic injuries.

Another general consideration is that the continuum of TBI care involves moving the patient across time, across functions of care (e.g., resuscitation, acute care, rehabilitation, disability evaluation, community/unit reintegration, chronic management), across geographic locations, providers, and treatment institutions of the Army, Navy, Air Force, VA, and private medical facilities. The continuum of care requires patient movement, caregiving, data recording, use of medical equipment, and medical supplies, as well as interactive communication among the patient, professional care providers, family members, social networks, and communities. Treatment or acute management of moderate or severe TBI is not as problematic as the long-term management of severe cases or the detection, tracking, and management of mTBI cases.

The objective of the analysis plan is to improve communication and coordination of care so that all TBI patients get the right care at the right time in the right place by the right provider. In an ideal system, patients, their families, and their care providers have access to necessary information and are fully informed and empowered to improve and accelerate recovery. The plan developed by the group provides a framework for achieving this objective.

The technical approach necessitates detailed mapping of the TBI care process showing the flow of patients through events (nodes), the time between nodes, and the resource requirements, including providers, information, and medical logistics (such as assessment tools and protocols) at each node. The care process is different for unrecognized mTBI cases, recognized mTBI cases, and moderate or severe TBI cases. Process maps would provide a basis for DOD to construct models to test whether changing events, changing the time required to reach a node, and/or changing the resources would improve the efficiency of care.

Information relevant to TBI needs that is currently being collected at various care levels in various medical facilities in the United States could be linked using current and evolving information technology (IT) integration techniques. DOD could create a virtual knowledge-management infrastructure for enterprise-level, institution-level, clinic-level, and patient-level knowledge and decision making. The data infrastructure could draw together traditional individual and aggregate data on medical care, as well as video recordings (patient interviews in clinical settings or remotely from home), and other archived data that could be used retrospectively to assess TBI exposure (black box video from military vehicle, personal, and environmental sensors). Another technical approach would be networking technologies using Web-based systems for patient-to-provider, family-to-provider, provider-to-provider, and patient/family-to-patient/family communication.

The structure of the integrated system is highly dependent on the intended uses of the data or information it provides. To begin designing a structure, one must first identify all of the databases that would be linked. The level of detail would be specified in the design depending on how the data were used. For example, a field medic assessing a soldier for TBI following the explosion of an improvised explosive device (IED) would have different data inputs and outputs than a neurosurgeon in the United States involved in comprehensive acute clinical management. At the enterprise level, data formats may be less specific and more aggregated than data at the medical-facility level used for the management of individual patients.

The structure would have portals for access by multiple users, such as TBI patients, families, providers, MHS managers, and the general public. It would also have to be accessible from remote locations via portable wireless devices. One glaring deficiency in the current data structure is the absence of a uniform taxonomy for TBI, especially mTBI, in internationally used medical diagnostic database systems, such as the International Classification of Disease (ICD). In addition, metadata requirements would be integral to the structure to provide a context and organization for the data.

Integrating multiple databases would require core (master) identifiers common to the component databases. The data outputs would depend on the intended use of the information. The data requirements would not be limited to numbers and text, but would include data from video, voice, sensors, and other dynamic behavior, that have been compressed and archived and could be used "on call."

Nor would data be limited to existing databases. New data would be generated, stored, and linked within the integrated system; examples might include a database that measures progress or outcomes when using new clinical protocols or a database that demonstrates whole-system performance using specified metrics.

A basic assumption for the information system is that different organizational entities involved in TBI care, such as military treatment facilities (MTFs), VA MTFs, and TRICARE-sponsored private facilities, would share data they had collected on behalf of their patients and on behalf of the accountable health system. Another assumption is that effective communication and coordination would entail much more than the application of information technologies. It would involve the commitment of resources, the establishment of compatible policies and procedures, and synchronous planning at the tactical (bedside), operational (hospital), and strategic (enterprise) levels involving all affected stakeholders. A final assumption related to information technologies is that existing database systems can be identified and integrated.

Any plan designed to improve communication and coordination in the comprehensive delivery of TBI care for military personnel would be subject to major constraints. Cultural differences among the Army, Navy, Marines, Air Force, VA, and the private health care systems, reflecting mainly differences in tradition and mission, are barriers to transparent interoperability. Even at the data level, there are numerous constraints. Data security imperatives and patient privacy concerns can limit, delay, or prohibit data sharing. Databases are often fragmented (e.g., the military services may have similar, but not identical, databases) and decentralized, making integration difficult. Data collection in the field, especially in a combat zone, is very challenging and is always subordinate to the military mission and operational safety. The capabilities of frontline combat military service members may be limited, especially if the data collection task is time-consuming or complex.

One of the main constraints on TBI data is that there is no universal diagnostic code for TBI. Instead, related codes are used, such as skull fracture [ICD-9 800-804; ICD-10 SO2], concussion [ICD-9 850-850; ICD-10 SO6], and late-effect codes. In addition, an incident that may cause an injury, such as an IED explosion 50 yards away that knocks down and dazes a soldier, may be difficult to characterize or document. Sometimes the exposure has delayed effects, and sometimes it goes unnoticed or unreported because no symptoms persist. Thus these data may not be captured. Even if symptoms persist, they may not be reported if the service member believes that reporting them will delay his or her return home.

Metrics must be relevant to all key stakeholders. For patients, metrics include ease of access, utility (value), and timeliness of the information they need to facilitate their recovery. For family members, metrics would be based on whether they have access to information about their son or daughter's condition, treatment plans (with appropriate consent), and available resources and whether they could communicate critical information to the provider team in a timely, effective way. For providers, metrics would focus on whether they could access critical information about their patients, including the results of diagnostic tests, laboratory tests, and records of treatments and medications.

Provider teams also need access to patients and their families in some longitudinal manner and access to the most current policies, protocols, procedures, and devices for optimal TBI care. For the enterprise manager, metrics would measure access to data generated from treatment sources so they could determine the number of cases by severity level, incidence trends, costs, and outcomes of TBI care. Also the MHS (enterprise) manager needs to know the quality and timeliness of the data inputs.

Execution of this plan would require eight major tasks. The first, fundamental step is mapping the care process for TBI to identify and organize time, place, person, data requirements, and decision points. Second, available databases would be indexed (located and characterized) to ensure that current data from multiple sources are properly structured and included. Third, strategies and tools would be identified for integrating databases (building an architecture). Once databases have been identified and characterized, information technology tools could be used to link them. A longer term goal would be to establish and link new databases. The analytical knowledge would be included in the data infrastructure to develop, distribute, and use information to inform future health care policy related to TBI.

The fourth task is to develop an interactive communication system, an open architecture system that can accommodate and connect the informational needs of patients, providers, families (caregivers), and the community using multiple paths, multiple modes, and multiple devices in every geographic space where patients reside. This task requires leveraging the Internet using Web 2.0 or further evolved systems (including Semantic Web services), a common portal that allows wireless and remote access, provides security protection, and has interactive capabilities. Assigning a lifetime e-mail address to each service member could facilitate longitudinal two-way communication and outreach.

The fifth task is to establish mechanisms for clearly documenting and tracking each TBI case, which may be especially challenging for mTBI cases. Developing uniform identifiers for TBI is essential, especially for mTBI. When available, archived video collected in vehicles and/or in high-traffic areas in the combat zone could be used to identify suspected cases of TBI early. This task also involves identifying a mechanism for collecting and archiving real-time and non-real-time multimedia data. Existing and novel telemedicine tools could be used to monitor and provide care for longer term TBI cases, especially in remote locations.

Sixth, a registry for TBI cases (somewhat like a tumor registry) would be established for long-term follow-up of individuals and the evaluation of population trends. Seventh, GPS technology would be used to identify the location of service members with TBI and to create density maps to inform decisions about the location of care and communication resources. Finally, a small portable device would be developed that can store and retrieve the medical information of service members (essentially, a hardened individual electronic health record). Data would be backed up wirelessly at the time of entry to prevent data loss in case the device is damaged or lost.

Major output elements to this plan would be a TBI-care process map; an integrated data system with an analytical knowledge center; a dedicated communication portal to meet the information needs of patients, providers, families (caregivers), and the community (public); and a TBI registry.

The major output of the analysis plan would be that the military service medical departments, VA, and TRICARE system leaders would develop and refine plans to improve the compatibility of care practices and data systems for TBI care. The director of the new Center for Psychological Health and Traumatic Brain Injury could play a leading role in implementing this plan. Clinical expertise could be drawn from the DVBIC staff, and engineering expertise would be available from sources internal to and external to DOD to facilitate process mapping and to provide advice on information technologies and OSE. Initial efforts would focus on feasibility analyses, cost and time estimates, and plans for pilot testing elements of the plan.

WORKING GROUP D: MEASURING AND FORECASTING THE DEMAND FOR TBI CARE[9]

The impetus for addressing the topic of demand for TBI care was to provide policy makers with data to make informed decisions about the

[9]See Appendix G for names of working group members.

resource requirements for meeting the future medical needs of service members and veterans with mTBI and to evaluate the cost effectiveness of TBI prevention and mitigation. Effective management of the resources—providers, facilities, equipment, and so on—required for the delivery of TBI care will necessitate an understanding of the current and projected need for care.

Three specific analysis objectives were assigned to Working Group D. First, based on historical data on mTBI, develop a statistical estimate of the number of mTBI cases in the population of military personnel who participated in OIF/OEF; second, develop a methodology for forecasting future mTBI cases in the military population, including new and previously undiagnosed cases; and third, develop the elements of, and a process for assessing the value of measures to prevent the occurrence of TBIs.

These three objectives are highly interrelated. The number of mTBIs among military personnel is not known with fidelity because it is generally accepted that there are unrecognized cases in the population. There are several possible reasons for this. First, the injury may be relatively subtle (changes in the ability to concentrate or changes in mood, for example) and therefore difficult to appreciate or acknowledge immediately. Second, injured persons may not report symptoms because they attribute them to other causes or because they wish to continue performing their duties. Third, symptoms may not develop immediately, or other injuries may mask mTBI-related deficits. Thus estimating the number of cases now and in the future is likely to require the same analysis approach and data. For this reason, the working group decided to address these two objectives together.

The initial challenge identified by the group was terminology. Databases, service members, medical providers, and other potential sources of information do not all use the same terms to describe mTBI injuries; in some cases the terms are used by more than one source but are defined differently. When the goal is to combine data to generate estimates, these differences create obvious problems. The group therefore assumed that a consistent terminology could be imposed on the data, that sufficient TBI data are available to make informed estimates, and that TBI diagnoses in the electronic medical records are accurate.

To achieve the analysis objectives, the group proposed that the military population be nominally characterized by branch of service, location in theater, and exposure risk. The input variables to the model—and thus the data required to perform analyses—could include the following:

- the number of mTBI cases diagnosed in OIF/OEF personnel at DOD and VA facilities and by other service providers
- the reported prevalence of TBI as a function of total casualties
- the distribution of reported TBI diagnoses by injury mechanism and severity
- the number of deployed service members
- the distribution of personnel by service and component[10]
- the distribution of personnel by theater (OIF versus OEF) and, in each theater, by combat or non-combat primary assignment
- the number of deployments[11] and duration of deployments[12]
- the rate of use of IEDs and other concussive devices by theater and region, if available

It was understood that these values would vary over time and that a determination would have to be made on the most appropriate timescales to use in constructing the models. The data would be derived from combat, medical, and government records, as well as historical data on OIF/OEF.

Thus the models would combine available data on diagnosed mTBIs with information on exposures and risks known to be associated with mTBIs to generate an estimated prevalence of TBIs[13] by mechanism (blast, assault, fall, vehicle accident, penetration wound, etc.) type[14] (mild, moderate, or severe), and time after the injury-producing

---

[10]Active-duty and reserve components (comprising both the "Reserve" and the National Guard) have different levels of training and may be assigned different missions, which in turn may affect potential exposures to TBI events.

[11]That is, is this the first time this unit has been there? the third time?

[12]That is, how long has the unit been there? Is this the first week? the last week of 15 months?

[13]Prevalence is defined as the total (diagnosed plus unrecognized or unreported) number of cases in a given population at a particular point in time.

[14]TBIs can also be classified by the cause of the injury—closed-head trauma versus penetrating wound. "Penetrating" is sometimes listed as a fourth TBI category, along with mild, moderate, and severe, because of the implications of this form of wound to the care of the patient. A workshop participant pointed out that phenomenology may also be important—whether an mTBI case presents as a headache, behavioral changes, and so on.

event.[15] These could be calculated as probability distributions to yield an understanding of the uncertainties associated with the estimates. The prevalence of mTBI, the ultimate goal of the exercise, will necessarily generate the broadest distributions, because the number of mTBIs is much less certain than the number of moderate and severe TBI cases.

The most critical constraint on constructing and implementing the models is the ability to develop a "good-enough" understanding of the progress of cases, from unrecognized to diagnosed, to be able to evaluate how well the surrogate exposure measures delineated above predict outcomes. A second constraint is the quality (availability and usability) of the data. Workshop participants indicated that the problems include the difficulty of identifying all relevant databases,[16] some of which may be service-specific or may be formatted in ways that are incompatible with other databases (e.g., coded in incongruent software applications or the same label used for different data). They noted, however, that efforts are under way to make databases more compatible and to create a central registry, although these efforts are still in the early stages. Data-mining techniques could be used in the short term to overcome some of these difficulties.

The value of the first (current population) model is that it will yield a more accurate picture of the impact of TBI on the military mission and its burden on the MHS and the VA health and benefits systems. The first model will also feed into the second (forecast) model, which can be used not only to project future demands on health and benefits systems, but also to predict the possible effect of various protective measures. These may include advancements in technology (e.g., more protective helmets), procedural changes (e.g., modifications to the assessment protocol), and new medical treatments (e.g., drugs that limit brain damage after injury). An understanding of how different injury mechanisms affect overall prevalence may also lead to more effective interventions and more efficient allocations of resources for the prevention of TBI injuries.

The working group identified several unresolved questions associated with the two prevalence models:

---

[15]Was the injury diagnosed at the time of the event? . . . post deployment? . . . after the person left the military?

[16]Relevant databases may include the Armed Forces Health Longitudinal Technology Application, Clinical Data Repository, TRANSCOM Regulating and Command & Control Evacuation System, and others.

- whether particular manifestations of mTBI are more likely to resolve quickly or can be successfully managed with relatively low-intensity treatment (e.g., short-term removal from the battlefield environment)
- the probability and determinants of delayed or sustained symptoms from a TBI
- the probability and determinants of long-term sequelae from a TBI
- the cumulative effect of multiple TBIs
- the extent to which information about TBI from other populations—explosive ordnance disposal personnel,[17] miners and building demolition workers,[18] and other combatants in other militaries—is relevant to the U.S. military experience in a combat environment

Answers to any of these questions could be incorporated into the models to refine the estimates.

The third objective—to develop elements of and a process for assessing the value of preventing TBI—is necessarily more complex than the first two, because prevention and injury mitigation can mean preventing TBI-causing events and/or providing better protective measures for mitigating injuries or keeping the health status of injured personnel from deteriorating by using best practices and new medical and other therapeutic technologies. Furthermore, assessing the value of prevention would entail not only the costs saved in lost-duty time, in impacts on other co-morbidities, in possible administrative separation, and in not having to provide health and social support services to the service member/veteran and family, but also changes in the injured person's quality of life, a critically important concept that is extremely difficult to quantify.

The intent behind the third objective was to develop a means of evaluating alternative TBI protection initiatives and, generally, to compare the utility of prevention and treatment approaches. Given

[17]The Defense Advanced Research Projects Agency (DARPA) is currently investigating this at the Marine Corps Air Station, Cherry Point, N.C., Explosive Ordnance Disposal School.

[18]Dr. Mouratidis, a member of the group, observed that one has to be careful about generalizing from other blast exposures, as her research indicates that exposure to a single IED blast may have a far greater effect than exposure to a large number of controlled blasts. COL Poropatich, also in the group, pointed out that one consideration is whether an individual is exposed to blast alone or blast plus the subsequent shockwave.

assessments of the various types and levels of prevention and injury mitigation, and using the previously discussed prevalence (and incidence) estimation models, it could be possible to estimate the new incidence of TBI by category (severity, mechanism, theater, etc.) as a result of the prevention, reduction, and/or amelioration of casualties attributable to the implementation of new technological and/or system protective initiatives. These new estimates of incidence and prevalence could then be used in the models of other task forces to determine the short- and long-term reductions in costs, resources, facilities, and locations of treatment for patients and their families in the DOD, VA, and civilian health care systems.

WORKING GROUP E: CAPACITY, ORGANIZATION, AND RESOURCE ALLOCATION FOR A TBI CARE SYSTEM[19]

TBI care in the MHS involves complex interactions at the tactical and strategic levels. In addition, numerous co-morbidities are associated with TBI, including mental health conditions and physical injuries. Addressing these issues entails an analysis of TBI-related capacity issues, organizational issues, and associated resource allocations within the MHS. Capacity issues involve requirements for providers, facilities, and equipment; organizational issues focus on assessing the cost effectiveness of TBI care, evaluating changes to the care system, and analyzing the impacts of multiple TBI care systems with different organizational structures. In addressing these issues, it is important to take into account system-wide interactions, as well as relevant TBI co-morbidities from a "systems" or "enterprise" perspective.

Group E was asked to develop a description of elements, processes, and activities that represent the dynamics of a complete episode of TBI care for injuries at all levels of severity; the description must include demand for TBI care, care processes (protocols), and resources. The purpose of the description is to inform the design of an approach to the development of a stand-alone model of the TBI care system or to enrich an existing enterprise-level health care delivery model that would include TBI system elements, care processes, and resources.

The overall output was envisioned to be a "TBI system" model or "enterprise-level" health care delivery model with the potential to address

[19]See Appendix G for names of working group members.

a broad spectrum of TBI capacity, organizational, and resource-allocation issues. The expectation is that, if the model(s) are properly structured, it might point the way to the prospective design of a TBI system of care. However, much of the work of Group E was predicated on the output of models or analysis plans developed by Groups A through D.

The second task of Group E was to outline the structure of a resource-allocation model or methodology for allocating scarce TBI care providers to meet the demands in theater and in CONUS for all TBI cases, no matter what the severity of the injuries. As an alternative to this approach, group members were asked to consider a system of assigning TBI patients to care providers. MHS care providers with specialized knowledge of TBI are often responsible for treating other diseases and are not geographically distributed in a way that enables them to provide efficient, effective care to existing and future TBI patients. Thus MHS needs a method for determining the best use of these scarce resources in the short term.

The purpose of an analysis of the resource-allocation issue was to address the shortage of MHS care providers with expertise in TBI treatment and management by providing a methodology for allocating scarce TBI-capable care providers to meet the needs of TBI patients in theater and in CONUS and to identify high-priority requirements for additional TBI care providers.

To achieve the overarching objective for Group E of developing a model for the allocation of resources in the MHS TBI care system, several important factors had to be considered—the first of which was deciding on a particular TBI model to follow. Since approximately 90 percent of TBIs experienced in theater are mild—and because moderate and severe TBI are both well understood and appropriately treated—the group decided the model for this analysis plan would be treatment of mTBI.

The questions to be addressed were: which TBI enterprise performance metrics matter and what affects them; which processes, resources, and organizations are necessary; which elements of the system are critical to success; and how scarce resources should be allocated. Important performance metrics included (1) coverage of care, (2) the percentage of mTBIs detected and treated appropriately, (3) outcomes in terms of patient safety and successful return to duty, (4) the percentage of individuals able to return to work, and (5) quality of life after an mTBI. Costs and trade-offs in the TBI care system were also considered.

To determine the necessary processes for the model, the group assessed how these processes are impacted by co-morbidities across the continuum of care, from disease prevention to patient/soldier rehabilitation and reintegration into society. The phenomena evaluated ranged from the physical to the social and included the need for a comprehensive monitoring process focused on awareness and education. The issues addressed included the typical sequence of events leading to the development of mTBI and differences between the presentation of symptoms and the observation of these symptoms over time. The consequences of the passage of time represent a large gap in our understanding of the development of mTBI; it is necessary that we understand when particular events associated with mTBI occur and whether the disease progresses if no treatment is administered. Resource allocation would also be assessed for all levels of TBI severity, and process measures related to outcome metrics would be analyzed.

Process characterizations in this analysis plan range from prevention to reintegration and include screening, diagnosis, treatment, and rehabilitation. Specific areas of focus were physical, cognitive, emotional, behavioral, and social factors, with monitoring of each to determine changes in an individual as a result of mTBI, the causes of the changes, how changes should be treated to maximize function, and how the individual can be helped to re-engage in work and social activities. Monitoring these factors requires awareness, education, pre-mission operational checks, subjective reports, physical assessments, and change management.

Critical issues for the analysis include the sequence of events before and after the occurrence of TBI, the presentation of symptoms and whether they are affected by the severity of the injury, observation of symptoms and subsequent effects of co-morbidities, and patient entrance into MHS. Other phenomena to be addressed by the model were (1) patient experiences compared with the experiences of others who are with the patient (family members, professional care providers) and (2) queues of individuals with progressing symptoms, including (3) the consequences of delayed treatment. The most salient difference between TBI and other diseases identified during the assessment were the difficulty of detecting mTBI, which requires treatment of symptoms that cross physical, cognitive, emotional, and psychological boundaries, rather than the fundamental causes of mTBI, which are poorly understood at present. Consequently, mTBI care requires multidisciplinary, coordinated care (more than 48 care programs have been developed). In the MHS this requires that each patient have a case manager (not a care provider) who serves as care coordinator. There is no evidence-based treatment for mTBI that accounts for multiple symptoms and co-morbidities. Thus an assessment of this care system must include analogues in the management of other complex diseases, such as diabetes and cancer.

To manage differences between mTBI and other diseases, variabilities in case management in MHS and VA at different locations must be considered. At present, there is one track for mTBI identified in theater, another for other injuries occurring in conjunction with mTBI, and a third track for mTBI recognized only after a period of time following the incident. Although the first point of contact is responsible for a patient regardless of the evolution of the patient's care, all patient care is ultimately the responsibility of specialists regardless of whether they have directly interacted with the patients.

The technical approach to this analysis plan involves the development of a description of elements, processes, and activities to represent the dynamics of a complete episode of mTBI care for use in modeling a TBI care system at the enterprise level. This includes an outline of the structure of a model or methodology to assist in planning for the allocation of scarce TBI care providers in theater and in CONUS. Thus the approach structure must define care paths that specify which functions are necessary and when branching and feedback paths occur and their associated criteria, and where most time is consumed in the system. Process maps would be developed, as necessary, and a model representation would be chosen with defined parameters (Figure 6-1).

Data requirements would be identified for estimating these parameters, and sensitivity analyses would be carried out to verify and test the model. Validation of the model would be accomplished through evaluation of the model relative to baseline data. Resource-allocation experiments would then be performed to assess, for example, the effectiveness and resource requirements for alternative disease-management protocols or the effectiveness of alternative distributions of a limited number of providers among multiple echelons of care.

Data requirements for this analysis plan are divided into four categories:

1. known aspects of mTBI with available data
2. known aspects of mTBI with no available data

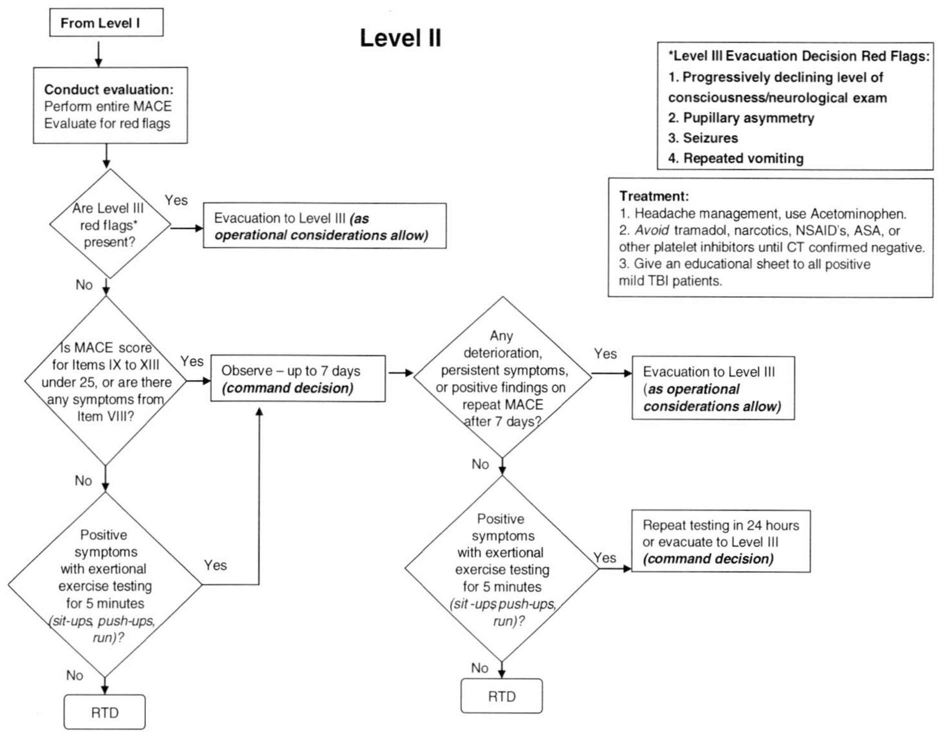

FIGURE 6-1 Sample mTBI care-process map for Level II treatment facilities. SOURCE: DVBIC Working Group, 2006.

3. aspects of mTBI with a recognized lack of understanding
4. aspects of mTBI yet to be identified

Present knowledge and currently available data are as follows: 11.2 percent of individuals surveyed in the military have reported mTBI; 50 percent of these individuals accessed care; and 10 to 20 percent of reports of mTBI are post-deployment (Labutta, 2008). In addition, both in theater and post-deployment reporting are currently designed to generate false positives. Ninety percent of care providers in the field use MACE,[20] and 85 percent of civilian non-blast patients recover from their injuries within three months.

Currently, we do not have substantiating data on the complete path of care in MHS, the effectiveness of rehabilitation therapy, or outcomes for those who return to work. Apparent issues identified with no supporting knowledge include the extent to which a TBI report indicates the presence of co-morbidities and the progression of TBI symptoms without treatment. Finally, there are still aspects of mTBI about which there is no information and no treatment.

A major assumption for this analysis plan is that the plans developed by Groups A through D, including the development of new TBI knowledge, the detection and screening of TBI conditions, the coordination and communication of TBI care, and the forecasting of the demand for care, will all accomplish their objectives. Nevertheless, work on implementing the Group E analysis plan can proceed in parallel with work on these other initiatives, as long as we allow for refinements in the model structure and data as additional information becomes available. Sensitivity analysis with early versions of the model can be used to help prioritize the need for additional information. In addition, standard modeling or computational engines involving standard representations must be used, and required data in existing information systems must be "capturable."

Critical constraints of modeling for this analysis plan include the lack of knowledge about care paths and results, the lack of baseline comparison data, and problems with determining the sample size because of the large number of care paths. The model would normally be calibrated against a baseline of performance of the system, but it is

---

[20]MACE stands for Military Acute Concussion Evaluation. For more information, see *http://www.dvbic.org/pdfs/DVBIC_instruction_brochure.pdf* (accessed September 29, 2008).

not currently clear what the baseline performance is or how one would assess it. Because of the large variety of care paths, many samples of patients with particular combinations of symptoms are too small to be statistically meaningful. Other constraints on the implementation of this plan include (1) acceptance of the model in the military culture, by individuals, and by the public and (2) resource constraints because moderate and severe TBI consume the majority of resources for TBI care and management, leaving few resources for mTBI care.

Performance, outcome, and utility metrics would include the input of a population from each of the 48 care program categories; an estimate of patient coverage (the percentage of individuals detected and treated appropriately and the percentage missed); and patient safety outcomes in terms of work-related activities, treatment accessibility (how far patients must travel for treatment), percentage returning to work, and quality of life. Other metrics are the costs and trade-offs of resource requirements (e.g., people, facilities, funding), as well as the people and time required to operate and maintain the model (including collecting data, estimating parameters, and conducting experiments). Overall, the usefulness of this analysis plan is that it could create interactive models capable of generating answers to various questions in only a few months.

A necessary task for the execution of this analysis plan is a value-stream analysis for chronic mTBI. Currently, hundreds of soldiers per month "screen" positive for self-reported blast injuries and symptoms, with roughly 50 percent of these diagnosed as symptoms related to mTBI. A value-stream analysis would involve the identification of a cohort of patients with symptoms three months after injury and the tracking of these individuals through the system for three months post-diagnosis by walking with them through the process and understanding what they experience. Then a larger set of patients would be tracked through the MHS information system and surveyed in comparison to all other patients.

This would enable researchers to observe what actually happens to patients at each step and to determine the value added of each step in terms of insight and treatment for particular outcomes. The path of each patient would be mapped through the system, with durations and branching frequencies. Completion of the model and essential outputs for users would take about six months.

Following the preliminary plan described above, a series of modeling spirals would have to be defined that would "spiral" through the development of process maps for "as is" and "to be" care of patients. A model representation, which would include a disease model showing the progression of mTBI, an organizational model showing how relevant organizations function, and a care management model showing how care is managed by these organizations, would then be selected. Parameters for the model representation would have to be defined, and data requirements would have to be identified for estimating these parameters. Once the model has been verified and validated through preliminary testing and evaluations, resource-allocation experiments could be performed.

The proposed model could be used to inform policy development, respond to congressional inquiries, and identify resource implications for policy changes in terms of people, training, education, and money. In addition, the model could further an understanding of the implications of health quality outcomes, determine optimum allocations of limited resources, and ascertain what is unknown or cannot yet be imagined in relation to emergent behaviors following mTBI. Users of the model would include DOD, Defense Centers of Excellence, MHS, and VA. Outputs of the model would be useful to policy makers, the Government Accountability Office and Inspector Generals, and authorizers and appropriators.

The expected outputs for the proposed model include (1) the identification of processes that need improvement and suggestions for improving them; (2) the identification of, and priorities for, processes that should be created; and (3) the identification of critical data that should be collected. Priorities for data collection include information identified through sensitivity analyses, processes that truly impact outcomes, and areas in which exact numbers are needed.

Implementation actions include acceptance of the model by DOD and the definition of a course of action, the development of a business case for return on investment of such an initiative, and the elaboration of how current modeling investments could yield much larger future benefits. These benefits would include both longer term returns in reduced workloads and lower costs for MHS and VA health care providers, as well as the shorter term benefits of informed research road maps; analytically defensible investment strategies; credible, compelling risk-management strategies; and prioritized data gathering for high-leverage information.

The requirements for this initiative are estimated to be six months for the value-stream analysis, including data collection and direct observation, and an additional six months for each subsequent spiral. The nature of the spirals will depend on the questions that emerge from previous spirals, assuming a collaborative effort and broader coverage of phenomena. Core competencies include proficiency in working with MHS, VA, mTBI, and modeling systems; the value-stream analysis will require contributions from social scientists. The total resources required would equal the value of the person years per month times six months times N + 1, where *N* represents the number of spirals.

SUMMARY

The challenges of the TBI workshop from the perspective of OSE are captured in Figure 6-2. Before OSE methods can be brought to bear on TBI care, there must be at least a preliminary understanding of the relationships between blast and concussive events and TBIs subject to the conditions of delivery or occurrence and the state of the soldier who is injured. The development of this understanding was addressed by Working Group A in an approach focused on the use of diagnostic and screening tools to establish pre- and post-event baselines, as well as conducting basic research on blast and concussive effects.

The current MHS TBI care delivery system must be better specified and understood for OSE tools and methods to be used effectively. The complex military health care delivery system includes facilities, medical logistical support, and personnel of the MHS, VA, and civilian health care systems, as well as the families of soldiers suffering from TBI and the soldiers themselves. One of the basic challenges associated with the delivery of care in this system is patient tracking and case management.

Working Group C suggested an approach to the development of an information system for tracking, monitoring, and cueing care delivery for all TBI patients. The approach focuses on combining the integration and augmentation of existing databases and a communication system that would ensure access to information and the dissemination of information to appropriate parties; the system architecture would be compatible with the care delivery system.

Finally, Group A noted that data available *outside* the military TBI domain, in the form of records of concussive and other closed-head

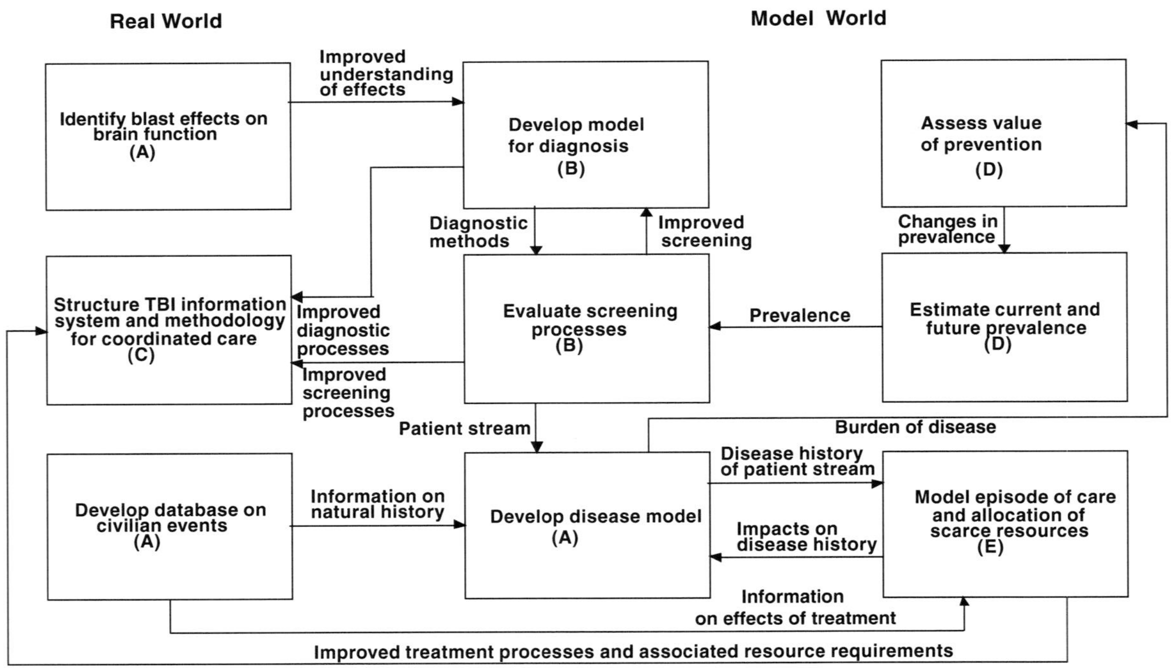

FIGURE 6-2 Interrelationships among suggestions for analysis plans developed by participants in Working Groups A through E.

injuries and cases of TBI in the civilian world, could contribute to the development of a more detailed understanding of the course of TBI and the effectiveness of alternative treatments. The group suggested an approach for integrating civilian and military data for use in the military environment.

The remaining blocks in Figure 6-2 show typical tasks, largely interrelated, that can be addressed by OSE techniques and methodologies. The development of a quantitative disease model is key to the evaluation of long-term demand for care, strategies for treatment, and the design (and costs) of a high-quality, efficient care delivery system. Working Group A developed an approach to modeling the course of TBI as a finite state-space stochastic process in which patients transition from state to state as a function of additional (non-TBI) trauma, treatment, and the long-term impact of trauma, with co-morbidities impacting the definition and occupancy times of any given state. Group A also developed a plan for using a survey methodology integrated with data mining to support the definition of states, the estimations of transition probabilities, and the distributions of occupancy times.

Parameters in the quantitative disease model include diagnosis and screening characteristics, state definitions, and state transitions. Working Group B outlined the development of a series of models that would quantitatively describe and evaluate current practices and then optimize the screening process. The analysis of data through data mining and the execution of surveys would be essential to the development of these models. Markov decision theory, Bayesian networks, influence diagrams, and simulation were cited as viable candidates for evaluating and designing TBI diagnostic and screening processes (see Appendix H).

To assess the capacity of MHS to treat TBI, one must first understand and be able to predict the "demand" on the system. Working Group D described the difficulties of estimating demand, which is complicated because, in many instances, the presence of TBI is not recognized or is not reported promptly in theater and can be missed in post-deployment interviews. The group observed that historical data could be analyzed to develop statistical estimates of the number of mTBIs in the current population of military personnel who have served or are serving in Iraq and Afghanistan. From there, the group outlined a methodology for forecasting future mTBI cases, taking into account conditions in the military theater, threats, and role-based exposures. The group also addressed the challenge of assessing the value of TBI prevention efforts.

The value of specific investments in prevention could be compared using methods of estimating the reduction of TBI incidence as well as the cost savings to the health care system and the reduced burden on soldiers and their families.

Given a quantitative disease model and an understanding of the effectiveness, availability, and costs of treatment, it is possible to design and evaluate care delivery from the perspective of individual patients, a patient population, and the entire health care enterprise. Working Group E designed an approach to the development of an enterprise-level health care delivery model with the goal of addressing, quantitatively, a broad spectrum of TBI treatment capacity, organizational and resource allocation issues to support decisions on policy, and design of the health care enterprise.

The suggestions for analysis plans developed by the five working groups indicate how OSE methods and tools could contribute to meeting the challenges of delivering effective, efficient, high-quality TBI care and management. Particular quantitative methods and models could provide insights into the design of diagnostic and screening processes, the delivery of care, the sizing of facilities, and the design of the overall health care delivery complex to meet current and future demands. As depicted in Figure 6-2, the challenges addressed by the OSE analysis plans are interrelated with the outputs of each analysis plan potentially providing important inputs to the development of one or more of the other plans. Collectively these illustrative plans address many of the TBI stakeholder issues identified during the planning phase of the workshop (Appendix B).

The suggestions and assumptions of the working groups identified underlying cross-cutting dependencies important to the development and application of OSE methods and tools to TBI care:

- an assumption that sufficient reliable data are available for the development of useful initial versions of all of the plans suggested by the working groups and that additional reliable data could be generated to improve and refine these approaches and provide more comprehensive and precise support to meet MHS needs
- a recognition of the need for standardized, detailed coding to record TBI symptoms, injuries, and (possibly) treatments more accurately

- the importance of data sharing and interoperability of data bases relevant to TBI diagnosis, treatment, measurement, and prediction
- the need for more extensive, detailed maps of care paths and processes and associated information, patient, provider, and material flows for TBI care within the MHS

The ideas and concepts introduced during the workshop may be helpful to DOD leaders working to refine and improve DOD's system of health care delivery, both at the individual patient-provider level and at the enterprise level. Applications of OSE concepts, tools, and methods have the potential to contribute to improvements in care, not only for TBI patients but for all patients cared for by MHS.

REFERENCES

Cope, D.N., N.H. Mayer, and L. Cervelli. 2005. Development of systems of care for persons with traumatic brain injury. Journal of Health Trauma Rehabilitation. Focus on Clinical Research and Practice 20(2): 128–142.

DVBIC Working Group (Defense and Veterans Brain Injury Center Working Group on the Acute Management of Mild Traumatic Brain Injury in Military Operational Settings). 2006. Clinical Practice Guideline and Recommendations 22 December 06. Available online at *http://dvbic.org/public_html/pdfs/clinical_practice_guideline_recommendations.pdf* (accessed August 7, 2008).

Guler, I., A. Tunca, and E. Gulbandilar. 2008. Detection of traumatic brain injuries using fuzzy logic algorithm. Expert Systems Applications 34(2): 1312–1317.

Labutta, R.J. 2008. Medical Aspects of Traumatic Brain Injury (TBI). Presentation at the Workshop on Harnessing Operational Systems Engineering to Improve Traumatic Brain Injury Care in the Military Health System, National Academies, Washington, D.C. June 11, 2008.

Sayer, N. 2006. Department of Veterans Affairs Quality Enhancement Research Initiative. 2006 Strategic Plan. Available online at *http://www.hsrd.minneapolis.med.va.gov/pdf/PT_Strategic Plan.pdf* (accessed September 29, 2008).

Appendixes

Appendix A

Biographical Information

NORMAN R. AUGUSTINE (NAE) (*co-chair*) retired in 1997 as chair and chief executive officer (CEO) of Lockheed Martin Corporation. Previously, he was chair and CEO of Martin Marietta Corporation. Upon his retirement, he joined the faculty of the Department of Mechanical and Aerospace Engineering at Princeton University. Earlier in his career, he had been under secretary of the Army and assistant director of defense research and engineering. Mr. Augustine has been chair of the National Academy of Engineering (NAE) and, for nine years, was chair of the American Red Cross. He has also been president of the American Institute of Aeronautics and Astronautics, chair of the Jackson Foundation for Military Medicine, a trustee of the Massachusetts Institute of Technology, a member of the U.S. Department of Homeland Security Advisory Council, chair of the Defense Science Board, a member of the boards of Black and Decker, Lockheed Martin, Procter and Gamble, and Phillips Petroleum, and chair of the Business Roundtable Task Force on Education. Mr. Augustine has received the National Medal of Technology and, five times, the highest award given by the U.S. Department of Defense, the Distinguished Service Medal. He has a B.S.E. and M.S.E. in aeronautical engineering, both from Princeton University, and has received 22 honorary degrees. He is the author or coauthor of four books, and was selected by *Who's Who in America* and the Library of Congress as one of "Fifty Great Americans" on the occasion of the 50th anniversary of *Who's Who*. He has traveled to more than 100 countries and has stood on both the North and South Poles.

DENIS CORTESE (*co-chair from April 2008*) is president and CEO of the Mayo Clinic in Rochester, Minnesota, chair of the Mayo Clinic Board of Governors, and a member of the Board of Trustees. A graduate of Temple University Medical School, he did his residency in internal medicine and pulmonary diseases at the Mayo Clinic. After serving in the U.S. Navy, he joined the staff of the Mayo Clinic in 1976 as professor of medicine. In 1993, he moved to the Mayo Clinic in Jacksonville, Florida, and from 1999 to 2002 he was CEO of the Mayo Clinic and chair of the Board of Directors of St. Luke's Hospital, both in Jacksonville. In 2003, he returned to Rochester. His major research interests have been in interventional bronchoscopy, including the appropriate administration of photodynamic therapy, endobronchial laser therapy, and endobronchial stents. Dr. Cortese is a member of the Institute of Medicine (IOM) and chair of the Roundtable on Evidence-Based Medicine and the Healthcare Leadership Council (2007–2009). He is also a former president of the International Photodynamic Association. Dr. Cortese is a member of the Harvard/Kennedy Healthcare Policy Group; Academia Nacional de Medicina (Mexico); the Royal College of Physicians (London); FRESH-Thinking (Focused Research on Efficient, Secure Healthcare); Advisory Board, World Community Grid; Chairs/Presidents/CEOs Council, American Medical Group Association; and Division on Engineering and Physical Science of the National Research Council. He received the Ellis Island Award in 2007.

JEROME H. GROSSMAN (*co-chair until his untimely death in April 2008*), was senior fellow and director of the Kennedy Health Care Delivery Project and chairman and CEO of Lion Gate Management Corporation. He was Chairman Emeritus of the New England Medical Center, adjunct professor of medicine at Tufts University School of Medicine, and Honorary Physician at Massachusetts General Hospital. Dr. Grossman chaired the Academic Medical Center Consortium, a cooperative of 12 leading academic medical centers that pioneered the use of health services research techniques to develop and implement effective and efficient health care delivery strategies. He was a member of IOM and a member of the IOM Committee on Quality of Health Care in America. He co-chaired the joint NAE/IOM study in 2007 on engineering and the health care delivery system. His career-long focus was on the use of information technology to improve health care delivery. He developed the first automated medical record system and founded Meditech, which serves more than 1,600 hospitals and integrated delivery systems, and Transition Systems Inc., which integrated clinical care algorithms with financial costs. He was a director/trustee of a number of organizations, including the Mayo Clinic Foundation, Penn Medicine (University of Pennsylvania Medical School and Health System), Stryker Corporation, Eureka Medical Inc., and the Committee for Economic Development.

SETH BONDER, president of the Bonder Group, is former chairman, CEO, and founder of Vector Research Inc. (VRI). Dr. Bonder has an international reputation in the field of policy and operations analysis for his development of new procedures that are used in the public and private sectors. Prior to starting VRI, he was a full-time faculty member in the Department of Industrial and Operations Engineering at the University of Michigan. During the past 40 years he has directed many force-planning, system-acquisition, readiness, and arms-control analyses for the Office of the Secretary of Defense, CINCS of the Unified Commands, Joint Chiefs of Staff, and U.S. Department of the Army. He has served as a consultant and advisor to the senior leadership of many of these agencies. His recent research has been focused on the growing problems of rising costs and reduced access to health care and the development of enterprise-level models and their use in prospective analyses of the reengineering of health care systems and management of chronic diseases. In 2000, Dr. Bonder proposed, and Congress authorized, a demonstration program by the Military Health System to test the utility of these models in evaluating alternative delivery policies, processes, budget development, and disease management protocols. Among the awards he has received are the Jacinto Steinhardt Memorial Award from the Military Applications Society of INFORMS and the Award for Patriotic Civilian Service from the Secretary of the Army. He has served on the U.S. Army Science Board, consulted for the Defense Science Board, and is a past president of the Military Operations Research Society and the Operations Research Society of America. Dr. Bonder received his Ph.D. in industrial engineering (operations research) from Ohio State University in 1965. He is a member of NAE.

PATRICIA FLATLEY BRENNAN is professor and chair, Department of Industrial and Systems Engineering, and the Lillian Moehlman-Bascom Professor of Nursing, School of Nursing, University of Wisconsin-Madison. Dr. Brennan is also affiliated with the university's Biomedical Engineering Center for Translational Research and the School of Education. Her fields of interest are health informatics, community health, information systems, computer-mediated clinical practice, and health services research. Her research has been focused on designing and evaluating home-care community computer systems for use by patients. Her work ranges from the development and evaluation of computer networks as a mechanism for delivering nursing care to homebound people and their caregivers to assessing the impact of patient-centered computer technology on the health outcomes of patients who have had coronary-artery bypass graft surgery. Her current projects are Project HealthDesign: Rethinking the Power and Potential of Personal Health Records (*www.projecthealthdesign.org*) and the healthsystems LAB (*http://healthsystems.engr.wisc.edu/*), which links the School of Nursing with the College of Engineering. These projects are exploring how individuals and families manage health information in their homes, studying the usability of secure e-mail in clinics, and developing information tools and resources to support self-care and health self-management. Dr. Brennan received her Ph.D. in industrial engineering in 1986 from the University of Wisconsin-Madison, her M.S.N. in nursing from the University of Pennsylvania in 1979, and her B.S.N. in nursing from the University of Delaware in 1975. She is a member of the Institute of Medicine, American Nurses Association, Wisconsin Nurses Association, Sigma Theta Tau, Alpha Mu, and Beta Eta. She is a fellow of the American College of Medical Informatics and American Academy of Nursing and associate editor of the *Journal of the American Medical Informatics Association.*

THOMAS F. BUDINGER is chair and professor of bioengineering and the Henry Miller Professor of Medical Research at the University of California at Berkeley; head of the Department of Nuclear Medicine and Functional Imaging at Lawrence Berkeley National Laboratory; and professor of radiology at the University of California-San Francisco. Dr. Budinger has made many contributions to a variety of scientific disciplines, ranging from polar ice exploration and space exploration medicine to the development of advanced technologies in the imaging sciences for the quantitative study of how the human heart and brain function in health and disease. He has received many grants from the National Institutes of Health (NIH), National Science Foundation, and U.S. Department of Energy. Dr. Budinger is a founder and past president of the Society of Magnetic Resonance, chair of the NIH Study Section on Diagnostic Radiology, and a member of the Scientific Review Committee for the Whitaker Foundation. He has chaired three NAE and IOM committees and is a member of both IOM and NAE; a fellow of the American Institute for Medical and Biological Engineering; and a member of the Society of Nuclear Medicine, American Association for the Advancement of Science, Sigma Xi, Society of Magnetic Resonance in Medicine, North American Society for Cardiac Radiology, and Alpha Omega Alpha. Other honors and awards include the American Nuclear Society Team Award for Special Achievement in Nuclear Technology for Medical Diagnostics, the Louise and Lionel Berman Foundation Award for Scientific Contributions in the Peaceful Uses of Atomic Energy, the Ernst Jung-Preis fur Medizin from Jung-Stiftung fur Wissenschaft und Forschung, the Paul C. Aebersold Award for Basic Science, the Hevesy Pioneer Award from the Society of Nuclear Medicine, a Distinguished Service Medal from the Society of Magnetic Resonance in Medicine, an NIH Merit Award for his contributions to Alzheimer's research, and an award for contributions to the NASA Apollo-Soyuz Project. Dr. Budinger is the author of more than 400 papers and three encyclopedia entries. His current interests include mathematics, chemistry, and instrumentation involved in imaging biological function and biomonitoring wireless technology for public health. He received a B.S. in chemistry from Regis College in 1954, an M.S. in physical oceanography from the University of Washington in 1957, an M.D. from the University of Colorado in 1964, and a Ph.D. in medical physics from the University of California at Berkeley in 1971.

BARRETT S. CALDWELL is an associate professor in the School of Industrial Engineering at Purdue University, where his research is on human-factors engineering approaches to how people get, share, and use information in complex environments. His work focuses on information flow, task coordination between and among team members, and team performance as affected by information and communications technology systems. His current research program is on advancing the state of the art in three specific domains of information system performance enhancement: (1) defining the quantitative characteristics of information flow in individual and team performance; (2) describing the effects of tasks, situations, and technologies on effective information exchange in organizations; and (3) making information more accessible for users working with computer system interfaces. Prof. Caldwell's research to date has focused on how information technology implementations and sociotechnical constraints affect the acceptance (in terms of timeliness and validity), use, and effectiveness (in terms of timeliness and the need for additional processing) of information flow and work processes at individual, group, and organizational levels. Approximately 10 projects since 1996 have been on information and document flow paths, time constraints affecting resource and provider coordination, human information needs, and adverse event paths in health care activities. Student projects, consulting, and NIH grants have included work in improvements in health care delivery in anesthesiology, blood services, cardiac recovery, clinical research management, laboratory, pharmacy, radiotherapy, rural health clinic operations, and surgical teams. Facilities in Ohio, Texas, and Wisconsin have been included in these studies. More than 10 peer-reviewed journal and conference papers, five student theses, and at least five seminars and presentations have been based on his research. A recent project on resource foraging and task coordination among teams of health care providers, funded by the Regenstrief Center for Healthcare Engineering, is a study of clinic operations in Indianapolis and East Chicago, Indiana. He is currently working on studies of workload transitions that affect inpatient pharmacy operations.

MICHAEL P. DINNEEN has been director of the Office of Strategy Management for the Military Health System since he retired from the U.S. Navy in January 2005. Following his medical training, he was a staff psychiatrist. He then transferred to the National Naval Medical Center, where he was first a director of residency training, then chairman of the Department of Psychiatry, and finally director of medical services. When Congress threatened to outsource all military mental health care in the National Capital Area, he developed and implemented a strategic plan to reduce psychiatric hospital beds from 200 to 60 while actually increasing the military's share of the mental health market. Subsequent changes resulted in an integrated training and service-delivery program that provided expanded child and adolescent services. At the same time, overall operating expenses were reduced by more than 30 percent.

Dr. Dinneen was also a special psychiatric consultant to the Secret Service and U.S. State Department, the attending physician to Congress, National Organization for Victim Assistance, and Office of the White House Physician. He developed his special expertise in psychological trauma and military psychiatry while leading Navy Special Psychiatric Rapid Intervention Teams for more than 10 years, directing Mental Health Services aboard the hospital ship USNS *Comfort* during Desert Shield/Desert Storm, and treating service members and their families. He has lectured internationally on traumatic stress, developed curricula in trauma psychiatry, and trained personnel for specialized wartime assignments. His publications on psychological trauma include original research on the effects of exposure to the stresses of deployment during Desert Shield and Desert Storm. In 2002, Dr. Dinneen became director of health care planning and TRICARE operations at the Navy Bureau of Medicine. He implemented a standard business planning process for the Navy's 38 medical treatment facilities and was responsible for the orderly transition to the new generation of TRICARE contracts. A diplomate of the American Board of Psychiatry and Neurology, Dr. Dinneen graduated from Harvard University (*cum laude*) and received both an M.D. and Ph.D. (in neurochemistry) from the Medical College of Virginia.

PAUL M. HORN was named New York University (NYU) Distinguished Scientist in Residence and NYU Stern Executive in Residence in September of 2007. Prior to his taking on the position at NYU, he was senior vice president and executive director of research of IBM Corporation, where he directed IBM's worldwide research program, which employed 3,200 technical workers at eight sites in five countries, and helped guide IBM's overall technical strategy. During his 28 years at IBM, Dr. Horn championed the translation of technology-based research into marketplace opportunities. Trained as a solid-state physicist who has held management positions in science, semiconductors, and storage, he has successfully applied advances in all of these disciplines to solving real-world technology problems. Dr. Horn's top priority as head of IBM's Research Division was to stimulate innovation and innovative business models and quickly bring those innovations to the marketplace to sustain and grow IBM's businesses and to create the new businesses of IBM's future. Prior to joining IBM in 1979, Dr. Horn was a professor of physics in the James Franck Institute and the Physics Department at the University of Chicago. He is a fellow of the American

Physical Society and was an Alfred P. Sloan Research Fellow from 1974 to 1978. He is a member of NAE, a former associate editor of *Physical Review Letters*, and the author of more than 85 scientific and technical papers. He has received numerous awards, including the 1988 Bertram Eugene Warren Award from the American Crystallographic Association, the 2000 Distinguished Leadership Award from the New York Hall of Science, the 2002 Hutchison Medal from the University of Rochester, and the 2002 Pake Prize from the American Physical Society. In 2003, Dr. Horn was named one of the top computing business leaders in the United States by *Scientific American* magazine. He is also a member of numerous professional committees, including the GAO board of advisors, the Gallaudet University Advisory Board, and the board of trustees of the Committee for Economic Development. He is also a member of the boards of trustees of Clarkson University, New York Polytechnic, and the New York Hall of Science and a member of the UC Berkeley Industrial Advisory Board. Dr. Horn graduated from Clarkson College of Technology and received his doctoral degree in physics from the University of Rochester in 1973.

COL (s) MICHAEL S. JAFFEE, M.D. (USAF) is national director, Defense and Veterans Brain Injury Center, Walter Reed Army Medical Center. Prior positions include director of the neurology program/OIC, neuropsychiatry, and principal investigator, Department of Defense and Department of Veterans Affairs Brain Injury Center, Wilford Hall Medical Center (WHMC), Lackland AFB, Texas; consultant on aerospace neurology, USAF School of Aerospace Medicine, Brooks City-Base, Texas; assistant clinical professor of neurology and psychiatry, University of Texas Health Science Center; and assistant professor of neurology/WHMC neurology clerkship coordinator, Uniformed Services University of the Health Sciences. Among the honors awarded him are the U.S. Surgeon General's Certificate of Appreciation (2003) for outstanding presentation at the national annual meeting of the Association of Military Surgeons of the United States and being chosen for the DOD/VA Clinical Guidelines Development Panel for Use of Narcotics in Chronic Pain (one of only four physicians selected from the Department of Defense to sit on the panel). He is a member of the American Psychiatric Association, American Academy of Neurology, Academy of Psychosomatic Medicine, and American Neuropsychiatric Association. Dr. Jaffee received his B.A./B.S. in economics in 1988 from the University of Pennsylvania, his M.D. in 1992 from the University of Virginia School of Medicine, and completed his residency in psychiatry and neurology in 1998 at WHMC, Lackland AFB-San Antonio.

CAPT WILLIAM P. NASH was commissioned in the U.S. Navy Medical Corps in 1978 after earning his M.D. from the University of Illinois College of Medicine in Chicago. His postgraduate training included an internship in surgery from 1978 to 1979 and a residency in psychiatry from 1982 to 1985, both at Naval Medical Center, San Diego. In addition to a tour as a naval flight surgeon, Dr. Nash served at Camp Pendleton, Portsmouth, Bremerton, and San Diego in several clinical, teaching, and leadership positions, including head of two Navy SPRINT (Special Psychiatric Rapid Intervention) teams and director of two Navy residency programs in psychiatry. He also served for five years as director of clinical services aboard the hospital ship USNS *MERCY* (T-AH 19) and six years as the Navy psychiatry specialty leader. In July 2001, Dr. Nash transferred to Marine Corps Base, Camp Pendleton, California, and, in January 2004, he joined the 1st Marine Division as part of the new Operational Stress Control and Readiness Program, which embeds mental health professionals directly with ground combat units. He deployed to Iraq with the ground combat element of I Marine Expeditionary Force (MEF) from August 2004 through March 2005, where he served as division psychiatrist and I MEF combat stress coordinator. Captain Nash was awarded a Bronze Star for the forward combat/operational stress management support and training he offered during that deployment. In October 2005, he was transferred to Headquarters, Marine Corps, to become the first combat/operational stress control coordinator for the Marine Corps, under the direction of the deputy commandant for manpower and reserve affairs. Along with Dr. Charles Figley, Dr. Nash co-edited, *Combat Stress Injury: Theory, Research, and Management* (Routledge, 2007).

ALEXANDER K. OMMAYA is chief officer, Translational Research, VA, where he facilitates collaborative projects with external groups in translational research, including pharmacovigilance and pharmacogenomics; develops translational research priorities and new administrative approaches for support; develops an integrated communication strategy and implementation plan for the Office of Research and Development; and is the representative to the VA Technology Assessment

Advisory Group. Previously, he was director of the Forum on Drug Discovery, Development, and Translation at the National Academies (2005–2006), where he established a multi-stakeholder group of leaders in the pharmaceutical and health insurance industries, government agencies that sponsor and regulate research and drug approval, and representatives of foundations, associations, and consumer advocacy groups. As director of the forum, he gathered support from 24 public and private entities and attracted key leaders; planned and implemented activities, such as research projects, workshops, and publications; and identified, developed, and implemented new collaborative projects, including the creation of template clinical trial agreements, the use of adaptive clinical trial designs, and drug development training. From May 2001 to December 2004, Dr. Ommaya was director of the IOM Clinical Research Roundtable. Before his work at the National Academies, he was manager of Blue Cross and Blue Shield of Florida; senior advisor, Agency for Health Care Policy and Research, U.S. Department of Health and Human Services; senior analyst, Henry M. Jackson Foundation, Walter Reed Army Medical Center; health policy fellow, U.S. Senate, Office of Senator George Mitchell; and senior biologist, Laboratory of Neuropsychology, National Institute of Mental Health. Among the awards he has received are the IOM Performance Award (2004); Blue Cross and Blue Shield Association Best Practices in Care and Primary Management (2000); and the AHCPR Administrator's Award for Group Achievements (1998). Dr. Ommaya received his Sc.D. in health policy and management from the Johns Hopkins University School of Public Health and his M.A. in biopsychology from Mount Holyoke College. He is a member of the VA Technology Assessment Advisory Group and a reviewer for the *Journal of the American Medical Association* and *Health Affairs.*

DAVID TRENT ORMAN is currently a civilian employee of the Department of the Army, where he heads the Post-Traumatic Stress Disorder-Traumatic Brain Injury/Behavioral Health (PTSD-TBI/BH) Integration Office. His role in the HQ U.S. Army Medical Command is to provide program integration/oversight in the behavioral health (BH) domains that impact soldiers suffering from PTSD/TBI and related BH syndromes. Dr. Orman brings 30 years of active-duty military and civilian administrative, educational, and research experience to his current position. Prior roles include eight years as the Army Surgeon General's psychiatry consultant, program director for psychiatric education at Tripler Army Medical Center, and associate program director for Texas A&M School of Medicine/Scott & White's psychiatry residency. He also was chief of psychiatry at Brook Army Medical Center and Darnall Army Community Hospital and division psychiatrist for the 1st Cavalry Division. He received his medical school education through the Uniformed Services University of the Health Sciences and completed his psychiatry residency at Walter Reed Army Medical Center, where he was chief resident in 1985. He is coauthor of numerous publications in the medical literature. Raised in the U.S. Air Force, Dr. Orman moved frequently with his retired E-9 father and DAFC mother.

COL RONALD POROPATICH is assigned to the U.S. Army Medical Research and Materiel Command (USAMRMC) at Fort Detrick, Maryland, where he is deputy director of the Telemedicine and Advanced Technology Research Center and the USAMRMC military liaison to the Department of Homeland Security. COL Poropatich is also the medical informatics consultant for the U.S. Army Surgeon General, where he works on implementation of information management/information technology solutions throughout the Army Medical Department. He is a former president and board member of the American Telemedicine Association and a practicing pulmonary/critical-care physician at Walter Reed Army Medical Center in Washington, D.C. He is also an associate editor of *Telemedicine and e-Health Journal.*

WILLIAM B. ROUSE is executive director of the Tennenbaum Institute and a professor in the College of Computing and School of Industrial and Systems Engineering at Georgia Institute of Technology. Dr. Rouse has written hundreds of articles and book chapters and is the author of many books; among the most recent are *People and Organizations: Explorations of Human-Centered Design* (Wiley, 2007), *Essential Challenges of Strategic Management* (Wiley, 2001), and the award-winning *Don't Jump to Solutions* (Jossey-Bass, 1998). He is editor of *Enterprise Transformation: Understanding and Enabling Fundamental Change* (Wiley, 2006), co-editor of *Organizational Simulation: From Modeling & Simulation to Games & Entertainment* (Wiley, 2005), co-editor of the best-selling *Handbook of Systems Engineering and Management* (Wiley, 1999), and editor of the eight-volume series *Human/Technology Interaction in Complex Systems* (Elsevier). Among his many advisory roles, he has served as chair of the National Research Council Committee on Human Factors, is a member of the U.S. Air Force Scientific Advisory Board, and a member of the DOD Senior Advisory Group on Modeling and Simulation. Dr. Rouse is a member of NAE, as well as a fellow of four professional societies—Institute of Electrical and Electronics Engineers, International Council on Systems Engineering, Institute for Operations Research and Management Science, and Human Factors and Ergonomics Society.

NINA A. SAYER is an investigator with the VA Health Services Research and Development Center for Excellence at the Minneapolis VAMC and the Center for Chronic Disease Outcomes Research and research director for the Polytrauma and Blast-Related Injuries Quality Enhancement Research Initiative (PT/BRI QUERI). In the latter capacity, she is responsible for directing the PT/BRI QUERI national research portfolio and implementing the results to improve outcomes for individuals who sustain polytrauma and blast-related injuries during their service in the Global War on Terror. From 1998 to 2002 she was co-director of training in psychology at the Minneapolis VAMC, where she was responsible for the American Psychological Association Ph.D. Psychology Internship Training Program. Her research on post-deployment health has included VA-funded studies of disability compensation, unmet needs for mental health services, community reintegration, and polytrauma. She is a member of the faculty at the University of Minnesota.

Appendix B

Issues Raised by Stakeholders about the Military Care of Patients with Traumatic Brain Injury

Caring for patients with traumatic brain injury (TBI) entails many challenges. Clearly, there are two TBI populations of concern, patients with moderate/severe TBI and patients with mild TBI (mTBI). Patients with moderate/severe TBI must navigate a complex and lengthy path through many military, Department of Veterans Affairs (VA), and civilian facilities where they receive acute care, rehabilitation, and chronic care by many providers and case managers. The much larger, but less obvious, population of service members who have had mild concussive injuries (mTBI) must be identified, treated acutely (if identified), triaged, tracked, and assessed for long-term outcomes and needs.

The following list of issues raised by stakeholders in the Military Health System (MHS) during the planning phase of the workshop addresses both mTBI care and moderate/severe TBI care. Following a meeting in February 2008, the workshop planning committee recategorized the issues on this long, though by no means comprehensive, list to correspond with five major challenges for TBI care: (1) the development of new TBI knowledge; (2) detection and screening of TBI conditions; (3) TBI care coordination and communication; (4) TBI care demand; and (5) TBI care system capacity, organization, and resource allocation. The committee then selected up to three issues from each of the five categories that were considered particularly well suited to illustrate the potential benefits of operational systems engineering (OSE) tools and techniques to TBI care management in the

Military Health System (MHS). The selected issues are explained in detail in a separate guidance document that was developed for the five working groups. (See Appendix C).

All of the issues raised by stakeholders (grouped by stages of medical care) are listed below to inform workshop participants of the wide range of issues that were considered by the planning committee.

Screening and Prevention

- There is no baseline cognitive screening tool
  - upon entry into the military
  - for pre-deployment screening
- There is no systematic follow-up assessment for those suspected or known to have a TBI—and no post-deployment screening.
- There should be a consistent, highly trained team to evaluate individuals with positive post-deployment screening.
- Personal protective equipment should be improved.
- Education of service members and their families about TBI symptoms and appropriate interventions is inadequate.

Diagnosis and Case Management

- No efficient, effective means of documenting trauma associated with a TBI (data on the type and severity of the exposure leading to the injury) has been developed.
- There is no gold-standard test for the presence or severity of TBI (no known biomarker, for example).
- Standard methods for assessing TBIs in the field or distinguishing when a concussion requires more intensive medical intervention are not used for all service members involved in a blast incident (MACE[1] algorithm tool is currently used).
- Gaps should be addressed in identification and subsequent treatment of soldiers with mTBI resulting from exposure to roadside bomb blasts—cases of mTBI may be overlooked.
- Differences between TBI symptoms caused by blast exposures and other traumatic exposures should be identified.

---

[1]MACE stands for Military Acute Concussion Evaluation. For more information, see *http://www.dvbic.org/pdfs/DVBIC_instruction_brochure.pdf.*

- DOD does not have a system-wide approach for properly identifying, managing, and monitoring individuals who sustain a TBI, particular mTBI.
- We need a better way of determining the incidence of brain injury and a secondary goal of tracking former soldiers who may have experienced brain injury but have since left active-duty military service.
- Missed diagnoses or premature return to duty may result in repeated concussions with long-term consequences.
- Only a limited network is in place to document and track service members from point of injury/diagnosis through post-military service care in the VA health system.
- There is no consistent disposition assessment for determining when/if a service member who has sustained a TBI should return to duty or at the time of discharge from service.
- Limited resources are available in theater (e.g., imaging).

Treatment

- Multidisciplinary treatment (physical, psychological, cognitive) is not available or used at all military/VA treatment facilities.
- No standardized care/treatment is provided at all military hospitals.
- Identified best practices are not uniformly implemented across the continuum of care for patients with all degrees of TBI. U.S. Department of Defense (DOD) does have a clinical practice guideline.
- Adequate treatment is lacking for patients in rural, non-urban, and underserved areas, who live too far from designated TBI centers or other VA treatment facilities to receive treatment from them.
- Problems in coordinating care between DOD and VA facilities—especially in transferring electronic records from DOD to the Deaprtment of Veterans Affairs—should be addressed.
- Communication and coordination among care providers at different levels of care and at different medical facilities is inefficient.
- There is no validated follow-up with appropriate clinical assessment techniques to recognize neurological and behavioral effects following acute injury other than MACE.

- We need more knowledgeable care providers with expertise specifically in TBI—many care providers currently have little or no experience in treating blast-related brain injuries.
- A challenge in treating mTBI is the co-morbidity of post-traumatic stress disorder (PTSD) and other psychiatric disorders, and some overlap of symptoms; PTSD associated with TBI may be different than non-TBI-related PTSD; understanding how TBI symptoms and psychiatric symptoms exacerbate and mediate one another.
- TBI may be harder to recognize in cases where there are no outward signs of injury.

Rehabilitation and Chronic Care

- Rehabilitation care is not standardized and is not always initiated when clinically indicated, nor has an optimal rehabilitation program been determined.
- Patients cared for in VA facilities who are still on active duty may complicate rehabilitation regimens.
- Excessive delays in establishing necessary services—mostly as a result of problems in transferring records from DOD to VA and/or in obtaining military health care benefits—should be eliminated.
- There are no objective measures for what constitutes recovery from a TBI, making the determination of fitness for return to duty problematic.
- Long-term case management is inadequate for service members/veterans who are impaired but not hospitalized; tracking of changes in health or mental status, quality of life, adherence to therapy/medication, etc. are spotty, at best.
- We need new programs to focus on coordinating care, such as a federal care-coordinator system (polytrauma) and Defense and Veterans Brain Injury Center (DVBIC) regional care-coordination systems.
- Support for families caring for TBI patients is insufficient or nonexistent.
- Ongoing efforts are being made with congressionally mandated Family Caregiver Curriculum.

- Role of community-level TBI-relevant services is not well defined or coordinated.

General Issues

- Additional TBI research is necessary along entire continuum of care, as well as a mechanism to collect data on the frequency, severity, care, and outcomes of TBI patients.
- The lack of data impairs analysis of the operation, cost, and effectiveness of TBI care; DOD and VA should systematically collect, code, retain, share, and analyze data in a way that respects the privacy of patients and the confidentiality of their information.
- The challenge in understanding, diagnosing, and treating military personnel is greater for mTBI than for moderate or severe TBI.
- No single medical discipline addresses TBI issues.

Appendix C

Operational Systems Engineering Applications Based on Issues Raised by TBI Stakeholders

A. Developing New Traumatic Brain Injury Knowledge

Issue A.1. Develop an approach to modeling the neuropathology and clinical effects of blast and concussive injuries on brain functions leading to mild, moderate, or severe TBI.

Purpose. Little is known about the phenomenology leading to TBI, particularly mild TBI (mTBI). This limits the objective diagnosis of TBI, effective management at the point of injury, and appropriate acute care. An understanding of the phenomenology would provide a basis for the development of effective treatment protocols.

Output. (1) A description of TBI phenomenology (e.g., are there useful "brain vital signs"). (2) An objective means of estimating the probability distribution of the severity of TBI as a function of blast and concussive effects on the brain that can be related to the origin of the blast.

Issue A.2. Develop an acute-to-chronic disease model of mTBI showing the evolution of disease states (symptoms?) over time for a population of mTBI patients, including both persons exposed to blast who are asymptomatic and persons who are overtly symptomatic.

Purpose. Little is known about the progression or disappearance of mTBI (related to blast injury) over long periods of time. This limits the effectiveness of triage, rehabilitation regimes, long-term chronic management of mTBI disease, and the best use of community services for mTBI patients.

Output. (1) An objective means of evaluating the efficacy of different intervention protocols for the near- and long-term care of mTBI patients. (2) A means of assessing the value of "early interventions."

B. Detection and Screening of mTBI Conditions

Issue B.1. Develop a model for the medical diagnosis/detection of mTBI based on current clinical experience of the events and processes leading to the onset and progression of the disease and on the questionnaires/testing of military personnel (e.g., Bayesian networks, influence networks).

Purpose. Not much objective knowledge is available about the onset and progression of mTBI that can assist in the detection and screening of mTBI patients. However, there is a lot of subjective information (e.g., experience in the medical community; neurological, cognitive and psychological testing; imaging; questionnaires) and mTBI incidence data that can be integrated as an interim diagnostic vehicle to assist in assessment, detection, and screening programs and in "return to duty" decisions.

Output. A means of estimating the probability that an individual soldier returning from the field has mTBI.

Issue B.2. Assuming the availability of a subjective mTBI diagnostic/detection methodology or another means of estimating mTBI detection probabilities, develop the structure and processes of a quality control program for screening the population of in-field and returning soldiers for mTBI.

Purpose. There is a need for better testing methods (cognitive, brain scans, other) that can be used (in combination with other information) to develop an effective, efficient screening process that appropriately considers Type I (sensitivity) and Type II (specificity) errors of the detection decision. In other words, we must improve the detection and identification of weak signals in the presence of substantial penalties of both false alarms (removing healthy soldiers from service) and misses (sending impaired soldiers back into harm's way). The signals may not be evident as physical symptoms.

Output. (1) A screening process and procedures for detecting mTBI in the population of soldiers returning from field operations. (2) An experience-based mTBI diagnostic/detection methodology and screening process as a means of assessing the utility of new testing methods (cognitive, brain scans, other).

C. TBI Care Coordination and Communication

Issue C.1. Develop the structure of a TBI information system to track, monitor, and cue care delivery for all TBI patients, no matter the severity of their injuries. The system should be useful for clinical monitoring and follow-up. In addition, it should be accessible to and cue all patients, patients' families, and other relevant providers in the MHS, VA, and civilian sector.

Purpose. DOD does not have a system-wide approach for tracking and monitoring TBI patients for effective management of their complete care. The coordination of care is poor between the MHS and VA systems, as well as among facilities and care providers at different levels and different medical facilities.

Output. A proactive information system that will facilitate the tracking, monitoring, cueing, coordination, communication, and scheduling of care for TBI patients from "cradle to cure," so that information flows and flows of care can be aligned to provide the most effective and timely status awareness and response capability for TBI patients.

Issue C.2. Develop a methodology for coordinating the delivery of services for TBI and related co-morbidities immediately following trauma exposure. The methodology should take into account the needs and preferences of patients and family members, as well as the resources (number and type of providers available, workload, etc.) and infrastructure of the relevant health care system.

Purpose. This methodology would improve the timeliness, coordination, and efficiency with which TBI care resources (care providers, equipment, material, supporting organizations, infrastructure) are brought to bear on the needs of TBI patients during the first two to three days after a critical event (initial injury or recognition of symptoms by caregivers or family members). The TBI care system must become more operationally responsive and better coordinated to improve both patient outcomes and to make efficient use of scarce resources.

Output: An operational model and/or a process methodology that can be used for the real-time allocation of TBI care resources in order to provide coordinated and responsive delivery of clinical services.

D. The Demand for TBI Care

Issue D.1. Based on historical data on known mTBI detections/patients and improvised explosive device (IED) incidents, develop a statistical estimate of mTBI in the population of military personnel who have participated in the Iraq and Afghanistan wars. The estimate should include the "shadow" population of mTBI patients.

Issue D.2. Develop a methodology to forecast the time stream of future TBI patients based on a projection of IED and other wartime blast phenomena in current and projected theaters of war. Specifically, based on historical and test data on various types of IEDs, develop a model to estimate the severity of concussive blast effects on individuals as a function of the input characteristics, such as blast sizes and types, proximity of blast to individuals, physical shielding and protections available, duration and number of blasts and/or incidents. The estimates of blast concussive effects should then be used, in conjunction with the results of Issues A.1 and/or B.1, to estimate the future demand for care of TBI patients.

Purpose. The effective management of resources available for TBI health care (providers, facilities, equipment, etc.) requires an understanding of the current and projected demand for care of TBI and mTBI patients.

Output. Estimates of current demand for TBI care and a methodology for estimating future demand for care of TBI and mTBI.

Issue D.3. Develop elements of, and a process for, assessing the "value" of TBI preventive methods (e.g., education, outreach, protective clothing and equipment, etc.) to DOD/VA, to potential TBI patients, and to their families/communities.

Purpose. The military should improve its efforts to prevent TBI.

Output. A means of assessing the value of alternative protection initiatives and a way to compare the costs and benefits of prevention and treatment.

E. TBI Care System Capacity, Organization, and Resource Allocation

Issue E.1. Develop a description of the elements, processes, and activities that represent the dynamics of a complete episode of TBI care at all levels of severity to include demand for TBI care, care processes (protocols), and care resources (providers, facilities, equipment). This description should be used for one of the purposes listed below:

1. To design an approach to develop a stand-alone model of the TBI care system.
   OR
2. To design an approach to improve an existing enterprise-level health care delivery model, including TBI system elements, care processes, resources, etc.

Purpose. TBI-related capacity issues (requirements for providers, facilities, equipment, etc.), organizational issues (assessment of the cost-effectiveness of the TBI care system, evaluation of changes to it, impact of multiple and different TBI medical systems), and associated resource allocations must be assessed. The TBI system of care involves many tactical-level and strategic-level interactions among elements, processes, activities, and organizations. In addition, there are significant co-morbidities between TBI and mental health conditions, as well as between TBI and physical injuries, diseases, and conditions. Therefore, these analyses should also consider endogenous interactions and relevant co-morbidities from a "systems/enterprise" perspective.

Output. A "TBI system" model or "enterprise level" health care delivery model that can address a broad spectrum of TBI capacity, organizational, and resource-allocation issues. If properly structured, the model(s) could be used to design prospectively a TBI system of care.

Issue E.2. Outline the structure of a (mathematical programming?) model/robust methodology to assist in planning for the allocation of scarce TBI care providers to meet the demand for care in theater and in the continental United States (CONUS) for all severity levels of TBI. (As an alternative, consider assigning TBI patients to specific care providers.)

Purpose. There is a shortage of care providers with expertise in TBI care. In addition, specialty providers may now have responsibilities for the treatment of other diseases and/or may not be geographically distributed to provide efficient care to the existing and projected population of TBI patients. Although it may be less than optimal from a systems perspective, the military needs a method to assist in determining the best use of these scarce resources in the near term.

Output. A methodology for allocating scarce TBI-capable care providers to meet the demand for care for in-theater and in-CONUS populations of TBI patients. The methodology will also help identify high-priority requirements for additional TBI care providers.

Appendix D

National Academy of Engineering/ Institute of Medicine Preliminary Information-Gathering Meeting: TBI Care System Mapping

December 20, 2007
11:00–16:00
Keck Center of the National Academies, Room 204
500 Fifth Street N.W.
Washington, D.C.

WORKING PLAN
(All times at the discretion of the co-chairs)

Goal:

Generate critical input for the February 2008 planning committee meeting for the Workshop on Harnessing Operational Systems Engineering to ImproveTraumatic Brain Injury Care in the Military Health System.

Means:

Bring together TBI care experts from the Military Health System (MHS) with different perspectives on the "system" of care delivery to work together with systems engineering experts to begin to characterize, model, and identify potentially high-yield opportunities for improving (and possibly redesigning) the MHS system of TBI care delivery.

Tasks:

- Task 1: Discuss the overall objectives of the project/workshop.

- Task 2: Reach a shared understanding of
  - the boundaries, structure, dynamics, and constraints of the current MHS TBI care system (clinical/work processes, work flow, patient flows, information, logistics flows, and the like), including the taxonomy of TBI and its implications and the major challenges facing patients, providers, and administrators with respect to system performance
  - the MHS TBI care system in the context of the universe of services for TBI-injured military personnel—the TRICARE system (including civilian care providers), the VA care and benefits system, private-sector health and insurance, federal Supplemental Security Income/Social Security Disability Insurance (SSI/SSDI) programs, et al.

- Task 3: Perform
  - initial assessment of the potential of systems engineering tools and technologies to maximize the effectiveness of MHS medical mission support, including the range of analysis, modeling, and design tools
  - initial identification of the MHS TBI care provision problems that can best be addressed using systems engineering principles

- Task 4: Identify potential case studies or scenarios that address challenges and opportunities for bringing systems engineering and information technologies to bear to improve TBI care.

National Academy of Engineering/Institute of Medicine TBI Care System Mapping Meeting

December 20, 2007
Room 204, Keck Center
500 Fifth Street, N.W., Washington, D.C.

ATTENDEES

Norman Augustine (co-chair) (NAE)
Retired Chairman and CEO
Lockheed Martin Corporation

Jerome Grossman, M.D. (co-chair) (IOM)
Senior Fellow and Director
Harvard Health Care Delivery Project

William Bograkos, M.D.
Colonel, Medical Corps, U.S. Army
Chief Warrior Transition Division
Clinical Operations, NARMC

Patricia Brennan, M.D., M.S.N. (IOM) [*by phone*]
Chair, Industrial and Systems Engineering
College of Engineering and
Moehlman-Bascom Professor
School of Nursing
University of Wisconsin-Madison

Paul Casinelli, M.D.
Brigadier General, U.S. Army

Lynda C. Davis, Ph.D.
Deputy Assistant Secretary of the Navy for Military Personnel Policy
U.S. Navy

Michael Dinneen, M.D.
Director, Office of Strategy Management
Military Health System
Office of the Assistant Secretary of Defense for Health Affairs

Randall Gay
Navy Lean Six Sigma (LSS)
Master Black Belt DASN (MPP)

Katherine Helmick, M.S.N.
Deputy Director
Clinical and Educational Affairs
Defense and Veterans Brain Injury Center

Michael Jaffee, M.D.
LTC., Medical Corps, USAF
Interim National Director
Defense and Veterans Brain Injury Center
Walter Reed Army Medical Center

Donald Jenkins, M.D. (*by phone*)
Colonel, Medical Corps, USAF
Director, Joint Theater Trauma System

Robert Labutta, M.D.
Colonel, Marine Corps, USA
Army Office of the Surgeon General

Leon Moores, M.D.
Colonel, Medical Corps, U.S. Army
Deputy Commander for Integration
National Naval Medical Center

William Nash, M.D.
Captain, Medical Corps, USN
Combat/Operational Stress Control Coordinator
Headquarters, Marine Corps (MR, M&RA)

David Orman. M.D. (*by phone*)
Chief, PTSD-TBI/BH Integration
HQ U.S. Army Medical Command

Ronald Poropatich, M.D.
Colonel, Medical Corps, U.S. Army
Deputy Director
Telemedicine and Advanced Technology Research Center

William Rouse. Ph.D. (NAE)
Executive Director and Professor
Tennenbaum Institute
Georgia Institute of Technology

Jamie Sinks
Nurse Practitioner
National Naval Medical Center

William J. Tanner
Commander, Marine Corp USN

Major Clifford Trott, Ph.D.
Chief Mental Health Officer
Office of the Chief Surgeon, NGB-ARS

NAE/IOM Staff

David Butler, Ph.D.
Senior Program Officer
IOM Medical Follow-Up Agency

Rick Erdtmann, M.D., M.P.H.
Director
IOM Board on Military and Veterans Health

Proctor Reid, Ph.D.
Director
NAE Program Office

Appendix E

Workshop Agenda

Harnessing Operational Systems Engineering to Improve Traumatic Brain Injury Care in the Military Health System
June 11–12, 2008
Keck Center of the National Academies, Room 201
500 Fifth Street, N.W.
Washington, DC 20001

Workshop Goal: To demonstrate the potential value of operational systems engineering (OSE) to improve the care of traumatic brain injury (TBI) in the Military Health System (MHS).

DAY 1

8:00 **Welcome and Introductions**

Norman Augustine, M.S.E., Lockheed Martin Corp (ret.) and Planning Committee Co-Chair (NAE)

Judith Salerno, M.D., M.P.H., Executive Officer, Institute of Medicine

S. Ward Casscells, M.D., Assistant Secretary of Defense for Health Affairs

SESSION I: BACKGROUND AND OVERVIEW

Session Objective: To provide a general background on the goals of the workshop, the current system of care for traumatic brain injury in the military, and general utility of operational systems engineering tools and techniques in other health care areas.

Session Moderator: Norman Augustine

8:20 **Traumatic Brain Injury Case Histories from the Military: "Survive, Thrive, Alive" Video**

9:00 **Medical Aspects of Traumatic Brain Injury**
Robert Labutta, COL, M.D., Medical Corps, US Army, TBI Senior Executive (Interim) Defense Center of Excellence for Psychological Health & Traumatic Brain Injury

10:15 **The "As Is" System for TBI Management in the Military Health System**
Michael S. Jaffee, COL. (s), M.D., (USAF) National Director, Defense and Veterans Brain Injury Center

11:15 **Operational Systems Engineering Examples from Other Health Areas**
William P. Pierskalla, Ph.D., Anderson Graduate School of Management University of California, Los Angeles (NAE)

SESSION II: WORKING GROUPS FORMULATE ANALYSIS PLANS FOR IDENTIFIED MHS TBI CARE SYSTEM ANALYSIS ISSUES

Session Objective: To demonstrate the kinds of approaches, methods, and information that can be developed by OSE practitioners to assist TBI care providers and managers.

After receiving guidance from the planning committee, five multidisciplinary working groups will convene to develop analysis plans for future OSE studies that could advance understanding of selected major challenges facing the TBI care delivery system.

11:45 **Target TBI Care System Analysis Issues for Working Groups**
Seth Bonder, Ph.D., The Bonder Group (NAE)

1:15 **Working Groups Session 1**

Each working group will be asked to develop brief analysis plans for future operational systems engineering studies that could be used to assist providers and managers address important challenges facing the TBI care delivery system. Each working group will work on pre-selected challenges in one of five major categories of TBI care analysis issues identified by the planning committee: (A) new TBI knowledge; (B) detection and screening of TBI conditions; (C) TBI care coordination and communication; (D) TBI care demand; and (E) TBI care system capacity, organization, and resource allocation.

Working Group Meeting Rooms:

Working Group A: Developing New TBI Knowledge [Keck 206]
Working Group B: Detection and Screening of TBI Conditions [Keck 207]
Working Group C: TBI Care Coordination and Communication [Keck 208]
Working Group D: TBI Care Demand [Keck 213]
Working Group E: TBI Care System Capacity, Organization, and Resource Allocation [Keck 201]

4:30 **Working Group Interim Reports and Discussion**
Moderator: Norman Augustine

5:30 **Adjourn to Reception and Working Dinner**

5:45 **Reception and Working Dinner**
Keck Center Atrium

Keynote: Insights from Vanderbilt's Journey toward System-Supported Practice

William Stead, M.D., Associate Vice Chancellor for Strategy/Transformation and Director of the Informatics Center at Vanderbilt University Medical Center (IOM)

DAY 2

8:00 **Welcome and Recap of the First Day**
Denis Cortese, M.D., Mayo Clinic and Planning Committee Co-Chair (IOM)

Norman Augustine, M.S.E., Lockheed Martin Corp (ret.) and Planning Committee Co-Chair (NAE)

SESSION II (Cont'd): WORKING GROUPS FORMULATE ANALYSIS PLANS FOR IDENTIFIED MHS TBI CARE SYSTEM ANALYSIS ISSUES

Session Objective: The five multidisciplinary working groups will reconvene to formulate "analysis plans" for two or more identified TBI care system analysis issues.

8:15 **Working Groups Session 2** (*in Breakout Rooms*)

Working Group Meeting Rooms:

Working Group A:	Developing New TBI Knowledge [Keck 206]
Working Group B:	Detection and Screening of TBI Conditions [Keck 205]
Working Group C:	TBI Care Coordination and Communication [Keck 208]
Working Group D:	TBI Care Demand [Keck 213]
Working Group E:	TBI Care System Capacity, Organization, and Resource Allocation [Keck 201]

SESSION III: REPORT OUT OF BREAKOUT GROUP DISCUSSIONS

1:00 **Working Group Reports and Discussion**
[Convene in Keck 201]
Moderator: Denis Cortese

SESSION IV: PANEL DISCUSSION ISSUES, OPPORTUNITIES, AND POTENTIAL NEXT STEPS

Session Objective: To identify the most valuable analysis plans set forth by the working groups that would, if implemented, significantly improve the care of TBI in the military.

Session Moderator: Denis Cortese

3:30 **Provider/Policy-Maker Perspectives on Working Group Outcomes: Panel and Moderated Discussion**
Michael S. Jaffee, COL. (s), M.D., (USAF) National Director, Defense and Veterans Brain Injury Center
Michael Dinneen, Director, Office of Strategy Management, Military Health System

4:30 **Concluding Summary Remarks and Adjournment**
Norman Augustine and Denis Cortese

Planning Committee:

Norman R. Augustine, M.S.E. (NAE/NAS), Co-Chair, Lockheed Martin Corporation (ret.)
Jerome H. Grossman, M.D. (IOM), Co-Chair (*November 2007 to April 2008*)
Denis Cortese, M.D. (IOM), Co-Chair, Mayo Clinic (*Beginning April 2008*)
Seth Bonder, Ph.D. (NAE), The Bonder Group
Patricia Flatley Brennan, Ph.D. (IOM), University of Wisconsin-Madison
Thomas F. Budinger, M.D., Ph.D. (IOM/NAE), University of California, Berkeley
Barrett S. Caldwell, Ph.D., Purdue University
Michael P. Dinneen, M.D., Ph.D., Military Health System
Paul M. Horn, Ph.D. (NAE), New York University
Michael S. Jaffee, COL (s), M.D., USAF, National Director, Defense and Veterans Brain Injury Center

William P. Nash, M.D., CAPT, Medical Corps, USN (ret.), USMC/USN Liaison to the Defense Center of Excellence for Psychological Health and TBI
Alexander K. Ommaya, Sc.D., Department of Veterans Affairs
David T. Orman, M.D., COL (ret.), HQ, U.S. Army MEDCOM
Ronald Poropatich, M.D., COL, Medical Corps, U.S. Army, Telemedicine and Advanced Technology Research Center (TATRC)
William B. Rouse, Ph.D. (NAE), Georgia Institute of Technology
Nina A. Sayer, Ph.D., LP, Department of Veterans Affairs Health Services Research & Development Center for Excellence

This workshop is dedicated to **Jerome H. Grossman, M.D.**, a long-time member, friend, and leader in the work of the National Academies. By nature and profession, Jerry was a bridge builder. He was the liaison between the Institute of Medicine and the National Academy of Engineering and the primary motivator and intellectual compass for this workshop and its focus on harnessing systems engineering tools, techniques, and knowledge to improve the quality of traumatic brain injury care in the Military Health System. He passed away suddenly on April 1, 2008.

Appendix F

Workshop Attendees

Tenley Albright
Director
Collaborative Initiatives at MIT

Priscilla Arriaga
Anderson-Commonweal Intern
NAE Program Office

Norman Augustine
Retired Chairman and CEO
Lockheed Martin Corporation

James Benneyan
Director, Quality and Productivity Laboratory
Northeastern University

Alfred Blumstein
University Professor & J. Erik Jonsson Professor of Urban Systems and Operations Research
Carnegie Mellon University

Seth Bonder
The Bonder Group

Thomas Budinger
Head, Department of Nuclear Medicine and Functional Imaging
E.O. Lawrence Berkeley National Laboratory

Jessica Buono
Research Associate
Policy and Global Affairs Division
The National Academies

David Butler
Senior Program Officer
Board on Military and Veterans Health
Institute of Medicine

Barrett Caldwell
Associate Professor and Director, Indiana Space Grant Consortium
Purdue University

S. Ward Casscells
Assistant Secretary for Health Affairs
Office of the Under Secretary of Defense for Personnel and Readiness
U.S. Department of Defense

Chanda Chhay
CASET Associates

W. Peter Cherry
Chief Analyst
Science Applications International Corporation

W. Dale Compton
Lillian M. Gilbreth Distinguished Professor of
Industrial Engineering, Emeritus
Purdue University

Denis Cortese
President and CEO
Mayo Clinic

Kenneth Curley
Chief Scientist
Neuroscience Portfolio Manager
Telemedicine and Advanced Technology Research Center

Michael Dinneen
Director, Office of Strategy Management
Military Health System

Rick Erdtmann
Director
Board on Military and Veterans Health and Medical
Follow-up Agency
Institute of Medicine

Louis French
Clinical Director
Defense and Veterans Brain Injury Center
Walter Reed Army Medical Center

Stephanie Guerlain
Principal Investigator
Medical Informatics and Training Program
University of Virginia

Paul Horn
Distinguished Scientist in Residence
New York University

Bernadine Hurst (Ctr)
Division of Science and Technology
Air Force Medical Support Agency

Michael Jaffee
COL (s), Medical Corps (USAF)
National Director
Defense and Veterans Brain Injury Center
Walter Reed Army Medical Center

Diane Kollar
Director, Industry and Government Relations
Tennenbaum Institute
Georgia Institute of Technology

Akhila Kosaraju
Office of the Under Secretary of Defense for Personnel and Readiness
U.S. Department of Defense

Robert Labutta
COL Marine Corps USA
Army Office of the Surgeon General

Lori Laraway
Captain (Select), Nurse Corp, U.S. Navy
National Naval Medical Center

Eva Lee
Associate Professor and Director for Operations Research in Medicine and HealthCare
School of Industrial and Systems Engineering
Georgia Institute of Technology

Henry Lew
Chief of Physical Medicine and Rehabilitation
VA Boston Healthcare System

David Maddox
General, U.S. Army (ret.)
Consultant
Arlington, Virginia

Randall McCafferty
LTC, U.S. Air Force
Chief of Medical Staff
59th Medical Operations Group
Lackland AFB

John McClellan
Corporal, U.S. Marine Corps

Michael McLoughlin
Deputy Executive
Biomedicine Business Area
Johns Hopkins University Applied Physics Laboratory

George Miller
Senior Analyst, Health Solutions Division
Altarum Institute

Maria Mouratidis, Psy.D.
Behavioral Health
National Naval Medical Center

William Nash, M.D.
CAPT, Medical Corps, USN (ret.)
USMC/USN Liaison to the Defense Center of Excellence for
Psychological Health and Traumatic Brain Injury

Alexander Ommaya
Chief Officer Translational Research
Department of Veterans Affairs

Hon Pak
LTC, Marine Corps USA
Telemedicine and Advanced Technology Research Center

Cynthia Petit
Division of Science and Technology
Air Force Medical Support Agency

William Pierskalla
Distinguished Professor Emeritus
Anderson Graduate School of Management
University of California at Los Angeles

Ronald Poropatich
COL, Medical Corps, U.S. Army
Deputy Director
Telemedicine and Advanced Technology Research Center

Proctor Reid
Director
NAE Program Office

William Rouse
Professor and Executive Director
Tennenbaum Institute
Georgia Institute of Technology

Julie Ryan
Lead Professor, Information Security Management
Assistant Professor, Engineering Management and Systems Engineering
George Washington University

Vinod Sahney
Senior Vice President and Chief Strategy Officer
Blue Cross and Blue Shield of Massachusetts

Judith Salerno
Executive Officer
Institute of Medicine

Nina Sayer
Investigator
Polytrauma and Blast-Related Injuries QUERI
Center for Chronic Disease Outcomes Research

Ariela Sofer
Professor and Chair
Systems Engineering and Operations Research
George Mason University

Barbara Springer
LTC, Medical Corps, U.S. Army
Chief of Physical Therapy
Walter Reed Army Medical Center

William Stead
McKesson Foundation Professor of Medicine and Biomedical Informatics
Associate Vice Chancellor for Strategy and Transformation
Vanderbilt University

Marco Steinberg
Associate Professor of Architecture
Stroke Pathways, Principal Investigator
Harvard Design School

Loree Sutton
BG, U.S. Army
Director
Defense Center of Excellence for Psychological Health and Traumatic Brain Injury

William Tanner
CDR, U.S. Navy

James Tien
Distinguished Professor and Dean
College of Engineering
University of Miami

Reha Uzsoy
Clifton A. Anderson Distinguished Professor
Edward P. Fitts Department of Industrial and Systems Engineering
North Carolina State University

Appendix G

Working Groups

Working Group A: Developing New TBI Knowledge
Thomas F. Budinger, *co-chair*
W. Peter Cherry, *co-chair*
W. Dale Compton, *rapporteur*
Alfred Blumstein
Kenneth C. Curly
Bernadine Hurst
Eva K. Lee
Michael P. McLoughlin
Alexander Ommaya
Staff: Jessica Buono and Pam McCray

Working Group B: Detection and Screening for TBI Conditions
James Benneyan, *chair*
Stephanie Guerlain, *rapporteur*
Louis French
Randall R. McCafferty
William Nash
Julie J.C.H. Ryan
Reha M. Uzsoy
Staff: Priscilla Arriaga

Working Group C: TBI Care Coordination and Communication
James M. Tien, *chair*
Barrett S. Caldwell, *rapporteur*
Tenley E. Albright
Paul M. Horn
Lori Laraway
Henry Lew
Hon Pak
Ronald Poropatich
Staff: Rick Erdtmann

Working Group D: TBI Care Demand
William P. Pierskalla, *chair*
David M. Maddox, *rapporteur*
Robert Labutta
Maria Mouratidis
Cynthia Petit
Nina A. Sayer
Ariela Sofer
Staff: David Butler

Working Group E: TBI Care System Capacity, Organization, and Resource Allocation
Vinod K. Sahney, *chair*
William B. Rouse, *rapporteur*
Michael P. Dinneen
Michael S. Jaffee
Diane Kollar
George Miller
Barbara A. Springer
Marco Steinberg
Staff: Proctor Reid

Appendix H

Definitions and Examples of Operational Systems Engineering Tools and Concepts

Agent-based models: See definition of simulation models below.

Bayesian networks are probabilistic graphical models that describe variables of interest and possible relationships (e.g., a patient's true medical status, field experience, test results, pre-existing status) and their probabilistic interdependencies. Bayesian networks encode probabilistic relationships among variables and account for circumstances in which data are missing and can be used to discover causal relationships (e.g., the relationship between symptoms and diseases). They are a good method for combining prior knowledge and newly collected data.

Bayesian decision trees: See definition of decision trees below.

Cellular automata (CA) models: See definition of simulation models below.

Cognitive task analysis is a method of identifying the cognitive demands on a user's cognitive resources (e.g., memory, attention, and decision making) from various aspects of system designs. This type of analysis is a way of looking at a system from the point of view of users and determining the thought processes that users follow to perform specific tasks. The information gained from such an analysis can help designers and users to focus on system features that users find hard to learn and to identify the points at which cognitive challenges might arise.

Decision-tree analysis is a tool that enumerates all possible outcomes of different choices in a given situation and computes the most likely result(s) of each. The purpose is to help a decision maker choose among decision options and to identify the strategy most likely to reach a particular goal. A decision tree takes the form of a graph with tree-like branches that shows all of the possible consequences of each decision option—including the probability, resource cost, and utility. In short, decision trees are visual and analytical decision-support tools for calculating the expected values (or utility) of competing alternatives.

Bayesian decision trees are a more advanced method that incorporates Bayesian networks into decision trees in order to account for uncertainties in the values and outcomes of decisions. Decision trees are also closely related to **influence diagrams** (see below).

Discrete-event models: See definition of simulation models below.

Fuzzy logic models are predictive or control models developed from fuzzy set theory that deal with reasoning and relationships that are approximate, approximately known, or estimated (rather than precise). Similar conceptually to probability theory (but different mathematically), fuzzy set theory is based on a graduated valuation of the degree of "membership" in various elements in a set (e.g., as a patient's screening test score increases, his or her degree of membership for a particular level of traumatic brain injury [TBI] severity rises or falls). The extent to which each element is true is described with a membership function valued on the (0, 1) interval. In fuzzy logic, the degree of truth of a statement ranges from 0 to 1 and is not constrained to the two truth values (e.g., does or does not have mild TBI [mTBI]). For example, fuzzy logic predictive models can assimilate "degrees of truth," or membership values, based on the results of screening tests to determine the most likely "state" (or status) of a patient.

Influence diagrams (also called decision networks) are compact graphical and mathematical representations of a decision situation (in a sense, they are generalizations of Bayesian networks) in probabilistic inference problems and decision-making problems. Influence diagrams are a tool for identifying and displaying the essential elements of a decision problem (e.g., decisions, uncertainties, and objectives) and how they influence each other.

Judgment models are a qualitative approach to making estimates based on consultation with one or more experts who have experience in the problem domain. For example, an expert-consensus mechanism, such as the Delphi technique, might be used to estimate the likelihood that a patient with a certain combination of presenting conditions does in fact have mTBI.

Markov chain models are stochastic processes in which a system (e.g., a patient or facility) transitions among a series of states (e.g., a patient being healthy, mildly sick, extremely sick, or dead; or a facility being empty, at half capacity, or full) and the Markovian (or "memoryless") property exists. The memoryless property means that the conditional probability of the system being in any given state in the future depends only upon its present state and is independent of any past states. (More advanced types of Markov chains can include several past states in the transition probabilities, but are memoryless beyond that amount of history.) Future states are reached by transitioning from one state to another with certain probabilities, rather than deterministically or with certainty. For example, given today's weather (state i at time t), tomorrow (time $t + 1$) it will be raining, cloudy, or clear j ($j = 1, 2, 3$) with defined transition probabilities $p_{i,j}$. At each step, the system may change from its current state to another state or remain in the same state, according to these transition probabilities. Changes between states are called transitions, and the probabilities associated with these changes are called transition probabilities.

Markov decision processes (MDPs) and Markov decision theory are extensions of Markov chains that provide a mathematical framework for modeling sequential decision making in situations in which outcomes are partly random (depending on actions or decisions by the decision maker). These models are often used to determine the optimal schedule of decisions, taking into account probabilistic events, demands, outcomes, and resource constraints. An MDP is a discrete-time stochastic control process characterized by a set of states (e.g., a patient's condition or the number of patients in a facility) and random (stochastic) future events. In each state, at discrete points in time, a decision maker can choose among several "control" actions (e.g., level of treatment, capacity expansion). For a current state, *s*, and an action, *a*, a state transition function, $P_a(s)$, determines the transition probabilities to each of the next possible states. The decision maker often earns a reward or penalty for each state that actually occurs. The state transitions of an MDP have the memoryless property described above (given that the state of the MDP at time t is known, transition probabilities to a new state at time $t + 1$ are independent of all previous states or actions). Note that the difference between Markov chains and MDPs is that MDPs include actions (allowing choice) and rewards (motivation).

Mixed-integer programming (MIP) models are mathematical optimization models that minimize or maximize a specified objective function, subject to a set of constraints (either linear or nonlinear); in MIP models, some of the decision variables are integers (e.g., the optimal number of facilities or medical personnel to locate in a given region). MIPs are heavily used in practice for solving problems in transportation and manufacturing, but they are also useful for some aspects of TBI care. For example, in a resource-location-allocation study, an MIP model was used to locate TBI treatment units in the Department of Veterans Affairs. The objective was to simultaneously determine optimal facility locations and the optimal assignment of patients to those facilities.

Monte Carlo models: See definition of simulation models below.

Partially observable Markov decision processes (POMDPs) are a variation of MDPs in which the current true state may not be known with certainty (e.g., a patient's true TBI status); instead, decisions (e.g., treatment, removal from field) are made based on current knowledge about the current state.

Sensitivity analysis is a general term for studying the impact on results of uncertainties in a model's logic or data, such as how different values of an independent variable or different processing steps will impact results. Typically, sensitivity analyses are conducted on uncertain values to explore how much the impact of a decision or policy will (or will not) change under different assumptions. Sensitivity analyses can be conducted on an ad hoc basis or more scientifically, such as by using the theory of experimental design. Sensitivity analyses can also be helpful for identifying the model assumptions that have the least (or most) impact on the results, which can be helpful when there are uncertainties in data. In particular, sensitivity analyses can help identify those data elements for which better estimates would be the most helpful and those that do not have to be specified very accurately in order to make a good decision. In this way, one can make better decisions about how to invest research time and money in the development, collection, and researching of data for a model.

Signal-detection theory (SDT) is used to analyze and optimize situations in which a decision is made that classifies ambiguous information (e.g., test results) into one of two categories (e.g., patient is sick or not) by trying to distinguish whether the observed result was created by the category of interest (called the *signal* in the SDT framework) or by random chance (called the *noise*). A common medical example is a blood test for a disease for which positive patients present with a range of numeric values and negative patients with a different range of values, but the ranges overlap—thus complicating the task of deciding whether a high result is a true "signal" or noise. SDT provides a mathematical framework for assessing such decisions—for quantifying the test's ability to discriminate and for determining the optimal threshold for calling a patient positive or negative.

Simulation models are computer models that emulate the logic of a process and use randomly generated data whenever a chance or random event (e.g., develops TBI, passes test, time durations) occurs in the model. Simulation models are very useful for studying the range of outcomes and most likely results of possible alternative process designs and courses of action. Such models often are used to analyze "what if" situations (e.g., what if we did something this way instead of that), or they can be used as part of an optimization computer program to find an overall optimal solution. Because of the flexibility and utility of computer simulations, they are widely used in operations research. Several types of simulation models might be helpful in modeling TBI:

- **Discrete-event models** are used to model the sequential/random flow of "things" (e.g., patients, personnel) through processes (e.g., the military field or the health care process), typically with the patient requiring various services that require various resources for various amounts of time. A discrete-event model is often used to assess system design and optimize flow, capacity, resource requirements, policies, and so on.

- **Monte Carlo models** are often used to analyze statistical problems that are otherwise "difficult" to solve. An example might be a series of integrated screening tests in which decisions are made after each screening (e.g., to conduct the next test, remove the individual from the field, or redeploy the individual); the analyst might be interested in determining the overall cost and accuracy (sensitivity, specificity) of a given process or protocol. For example, Monte Carlo models have been used to analyze and optimize cancer screening decision processes.
- **Agent-based models** are based on the idea of "agents" that represent each autonomous or semi-autonomous decision maker who chooses his or her next action based on the current status of the surrounding environment. This type of model is often used to model wartime theaters and other engagement activities, but for TBI it might be used to model medical decision making.
- **Cellular automata (CA) models** are used to model geographic movements in situations where the probability that a number of "things" (e.g., soldiers, patients) will move from their current grid locations to adjoining cells is dictated by the activities and state of affairs around them. CA models are widely used to model the spread of disease, species migration, forest fires, and other such events. For TBI, a CA model might be used to model the general geographic dispersal and flow of patients through different medical states or physical locations.

Value-stream analysis (VSA) is a tool used to evaluate all of the specific actions involved in a process, determine the relative value added of each action, and identify waste. VSA is often used to eliminate wasteful steps and create efficient processes comprising only value-added activities that maximize performance. With this type of analysis, one can separate activities that contribute to value creation from activities that create waste and then identify opportunities for improvement.